T0251116

INTERNATIONAL EQUITY AND GLOBAL ENVIRONMENTAL POLITICS

For

MARJORIE W. HARRIS

In memory of

GORDON F. "LEFTY" HARRIS

And his old New Hampshire appreciation for all things natural and wild

International Equity and Global Environmental Politics

Power and principles in U.S. foreign policy

PAUL G. HARRIS
London Guildhall University and Lingnan University

LONDON AND NEW YORK

First published 2001 by Ashgate Publishing

Reissued 2018 by Routledge
2 Park Square, Milton Park, Abingdon, Oxon OX14 4RN
711 Third Avenue, New York, NY 10017, USA

Routledge is an imprint of the Taylor & Francis Group, an informa business

Publisher's Note
The publisher has gone to great lengths to ensure the quality of this reprint but points out that some imperfections in the original copies may be apparent.

Disclaimer
The publisher has made every effort to trace copyright holders and welcomes correspondence from those they have been unable to contact.

A Library of Congress record exists under LC control number: 2001091619

ISBN 13: 978-1-138-73566-8 (hbk)
ISBN 13: 978-1-315-18648-1 (ebk)

Contents

Foreword

It is scarcely necessary to be a student of international affairs to see that preponderant power gives the society that enjoys it unique purchase in international relations. With the demise of Soviet power, the United States of America has been left, alone by a wide distance, at the head of the league table of world powers. If its position was contested up to 1990, we can nevertheless say with some certainty that the period beginning in 1940 is the American century. Despite the aspirations of some, Europe will have to do a great deal to reorganize itself before it can come to rival the United States; and both China and India have a long way to go before they can contest the top spot, even if at least one of them may do so within a generation.

America's preponderance is not just a question of economic and military strength. It is also a question of cultural strength, both at the popular and elite levels. The popular media across the world are dominated by American material, which shapes the minds and lifestyles of coming generations. For better or worse, the young in particular identify with American popular music and a range of American artifacts and foods, for example. This appears to be so even where there are strong forces of domestic cultural resistance. At the level of elites, there are aspirations to American lifestyles and consumption patterns, as well as subscription to patterns of thought and philosophies of political economy and institutional management that owe much to the Atlantic tradition. As at many times in history, a particular national culture is the standard-setter and supplier of the lingua franca for an international elite. In our times, much to the frustration of Europeans, Indians, Chinese and Japanese, and particularly of the defenders of French civilization, that leading culture is undoubtedly American.

The influence of that culture is abundantly evident in international environmental policy. Recent books edited by Paul G. Harris comprehensively demonstrate how the international community has often had to dance to the American tune in the last 20 years, not only on ozone but also on climate change and other environmental issues. Since global climate change policy cannot be effective if the United States is left outside it, the United States has been able to deploy its power and influence to shape agreements in ways acceptable to it. The same kind of processes can been seen in other areas, especially those which the American people and their tribunes perceive as being significant for the US economy (for example, biodiversity, agricultural technology and trade issues). Beyond that, a wider intellectual influence can be discerned, touching on basic approaches to environmental policy and the kind of policy tools to which preference should be given. This is reinforced and mediated through the Organization for Economic Co-operation and Development, the World Bank, and other international institutions.

So Paul G. Harris is mining a very rich vein through his expertise in U.S. environmental policy. In this volume, he looks closely at the issue of international equity. If climate change is looked at through the lens of ethics, questions of distributional impact—of who will bear the burdens of change, of who will be better off and who worse off—loom very large. These questions must be of concern to local, national and international communities for purely prudential reasons: sharp changes in wealth, the exacerbation of extreme poverty, and severe food insecurity can all be expected to breed tension and armed conflict. But bound up in such prudential considerations are ethical questions: the sense of injustice is, for example, in itself a powerful motivator of conflict. Moreover, in democracies, issues of morality, moralism, emotion, international reputation and pride often muddy the waters, as more traditionally educated diplomats are sometimes inclined to argue, of national self-interest in foreign policy. This is particularly so in today's global democracy facilitated by the electronic media. And the ethical, even the moralistic, dimension has been significant in US foreign policy from the very foundation of the republic.

More widely, the ethical dimension of environmental policy is itself of great, but often completely un-remarked, significance: almost any environmental goal, and any environmental tool or set of tools to attain the goal, raises the question of who has what and who will have what, and therefore the question who *ought* to have what. This tends to be well recognized with respect to the environmental externalities imposed by capitalist firms. However, the universality of the issue—the externalities that *each person* imposes—tends to be ignored. When the dimension of human rights and the rights of the rest of nature are raised, new layers of complexity are introduced. Penetrating still further into environmental ethics, there are layers of complexity, too, in the ethics of process.

Paul G. Harris is to be applauded for his happy foresight in bringing together the issues of U.S. environmental foreign policy and environmental ethics, in particular the question of equity. I am particularly happy that a period as a Visiting Fellow at the Oxford Centre for the Environment, Ethics and Society (OCEES) at Mansfield College, Oxford

University, gave impulsion to the interests demonstrated in this book. One of our missions at OCEES is to help policymakers and citizens, as well as academics, think more consciously and carefully about the relationships between ethics, environmental change, and how we live. That is what this book does. It is a welcome addition to thought about these issues, and I am grateful that OCEES is associated with it.

Neil Summerton, Director
Oxford Centre for the Environment, Ethics and Society
Mansfield College, Oxford University
March 2001

Preface

In this book, I argue that the U.S. government, particularly under President Bill Clinton, came to regard *international environmental equity*, defined as a fair and just distribution of the benefits, burdens, and decision making authority associated with international environmental relations, as an important feature of its foreign policy. I also acknowledge that U.S. efforts to implement this policy were quite limited. Indeed, while equity has become an important feature of global environmental politics, most of the economically developed countries have done too little to act on it. I wanted to explain this limited acceptance of equity in global environmental politics generally, and in U.S. foreign policy in particular. I wanted to know why the United States started to move in the "right" direction, and I wanted to understand why it did not do more to promote equity in international environmental policy making. Hence this book.

Questions about international environmental equity are more than academic. Serious efforts to bring equity into global environmental politics can lead to policies and practices that help protect the natural environment on which all countries and all people depend. Additionally, because it must be manifested by more aid to the world's poor countries, more serious consideration of equity in global environmental politics is likely to lead to less poverty and suffering on a potentially grand scale. By acting to promote international environmental equity, the United States and other developed countries can promote their national interests *and* the interests of people in the developing world. This would be a very good thing.

Some readers may misinterpret what follows as a strong assertion that the United States has embraced international equity and followed through with commendable steps toward its implementation. That is not what I want to say. Looking at last year's sixth conference of the parties (COP-6) of the Framework Convention on Climate Change (FCCC), in which the United States grappled with European and many developing countries over how it would meet its obligations under the FCCC's Kyoto Protocol, one can find plenty of legitimate reasons to criticize the United States. It failed to agree to concrete measures designed to limit its greenhouse gas emissions, despite its vastly disproportionate share of them. In short, it failed to lead by example. Nevertheless, I wish to argue that international equity is now an important consideration and component in U.S. foreign policy, in large part because of global environmental changes.

Indeed, COP-6 showed this to be the case; discussions of how to *implement* international environmental equity were perhaps the most distinctive feature of that conference. The United States argued about *how* it would implement provisions of the Kyoto Protocol, not *whether* it would do so.

This is a frequently overlooked shift in U.S. policy, and an essential precursor to the much more robust actions the United States must take if it is to lead the world in protecting the global environment. Sometimes international environmental equity makes it to the front burner of U.S. foreign policy, but most of the time it will of course be viewed as less important than other issues, notably issues like trade and military threats to U.S. interests and allies. However, this may change as environmental conditions grow worse. I think that in the decades to come international environmental equity will become much more prominent in global environmental politics *and* U.S. foreign policy.

This book contains three parts. In the first part, I introduce the concept of international environmental equity, try to define its meanings, and show how it has started to permeate global environmental politics and U.S. foreign policy. The second part is devoted to understanding and explaining international environmental equity in U.S. foreign policy. Eschewing simple explanations, I argue that the U.S. response to the emerging international consensus on environmental equity has been belated and imperfect, resulting from (1) concerns about the impact of environmental changes on U.S. national interests, (2) the complexity of America's pluralistic policy-making process, and (3) the subtle—but sometimes important—influence of the principle of international equity *per se*. In the final part of the book I discuss the implications for the United States and the world of embracing—or failing to embrace—international equity as a core objective of global environmental policy.

American leadership on international environmental equity is by no means certain, and recent events suggest that it may remain elusive. As this book goes to press, the new administration of George W. Bush is consolidating its power in Washington. In one of his first and more controversial acts, the new president reversed his campaign pledge to place new controls on U.S. carbon dioxide emissions, and he parroted climate skeptics who question the dangers posed by global warming. By all indications, the same forces shaping his father's policies on climate change were at work. The message of the president's "flip-flop" is one of the themes permeating this book: Political pluralism makes it extraordinarily difficult for the United States to take on its responsibilities in combating climate change and other environmental problems. The forces resisting U.S. environmental leadership are truly monumental, and the new administration is clearly very sensitive to them. Nevertheless, the new Bush administration may find it difficult to backtrack too far, and if it does so other countries may surprise us by finding the political will to act without the United States. Over time, this may stimulate a more equitable response from the United States.

This is the third book from the Project on Environmental Change and Foreign Policy. It almost goes without saying that foreign policy—the objectives that officials of national governments desire to achieve, the values and principles underlying those objectives, and the methods by which they are sought—can play an important, often vital, role in determining whether countries join international efforts to address environmental problems and act to actually address them. Yet, the processes of foreign policy have received relatively little systematic attention by scholars of environmental policy and politics. The project seeks to start remedying this situation. Like this book, the project's first two books, *Climate Change and American Foreign Policy* (St. Martin's Press) and *The Environment, International Relations, and U.S. Foreign Policy* (Georgetown University Press), focused on the environment in the context of international relations generally and U.S. foreign policy in particular. The next books from the project will be devoted to environmental change and foreign policy in the Asia-Pacific region. I hope that all of these books will be of interest to students and scholars, but I also hope they will be useful for policymakers and stakeholders grappling with the hard questions and choices surrounding changes to the natural environment.

I first exercised many of the ideas presented here in journal articles and chapters in other books. I wish to thank the persons who commented on those earlier papers, and the publishers for allowing me to use those ideas in portions of this book. I am grateful to Tom Trout, who was in some ways the catalyst for this book, and who has served as a dignified role model for me. I gratefully acknowledge the guidance given to me on this book by Seyom Brown, particularly with regard to its normative aspects. I will always appreciate his mentoring, and I will continue my feeble attempts to emulate him. In addition to Seyom Brown, Bob Art and Henry Shue gave me helpful comments during the early stages of this book, as did several readers who remain anonymous. I greatly appreciate their help. I would like to acknowledge the assistance of Joe Ng, who helped prepare the final manuscript, and Franceska van Dijk, who kindly read through it. I am especially grateful to Patricia Siplon for her advice and friendship over the years. She is a truly caring person who, through her intellect and good works, greatly inspires friends and students alike. Finally, I wish to thank the Oxford Centre for the Environment, Ethics and Society, and particularly its director, Neil Summerton, whose thoughtful foreword is, I trust, a propitious beginning to this book.

Paul G. Harris
London, England
March 2001

PART I

CONSIDERATIONS OF EQUITY IN INTERNATIONAL ENVIRONMENTAL POLITICS

1 Introduction: Environment, Equity, and U.S. Foreign Policy

During a press briefing in mid-1993, State Department counselor Timothy Wirth (soon to be under secretary of state for global affairs) declared that the Clinton administration was determined to reestablish the United States as the world's environmental leader: "the United States once again resum[es] the leadership that the world expects of us. [S]ee the changes that we have made related to environmental policy coming out of the disastrous events in Rio just one year ago at the UNCED [United Nations Conference on Environment and Development]. . . . Just a year ago, the United States was viewed as a country not fulfilling its responsibilities, and now we are, on these most difficult issues, once again out in the lead."[1]

That same year, Vice President Al Gore, speaking before the United Nations Commission on Sustainable Development, said that the United States and other developed countries "have a disproportionate impact on the global environment. We have less than a quarter of the world's population, but we use three-quarters of the world's raw materials and create three-quarters of all solid waste. One way to put it is this: A child born in the United States will have 30 times more impact on the earth's environment during his or her lifetime than a child born in India. The affluent of the world have a responsibility to deal with their disproportionate impact."[2]

In 1994 President Bill Clinton told the National Academy of Sciences, "If you look at the rate at which natural resources are disappearing and you look at the rate at which the gap between rich and poor is growing, if you look at the fact that the world's population has doubled [in only 40 years], it is clear that we need a comprehensive approach to the world's future. We put it under the buzzword of sustainable development, I guess, but there is no way that we can approach tomorrow unless we are at least mindful of our common responsibilities in all these areas. . . . already one-third of [the world's] children are hungry, two of every five people on Earth lack basic sanitation, and large parts of the world exist with only one doctor for every 35,000 or 40,000 people. Reversing these realities will require innovation and commitment and a determination to do what can be done over a long period of time. . . ."[3]

Testifying before the Senate Foreign Relations Committee in early 1995, Secretary of State Warren Christopher said that the United States can no longer escape the consequences of environmental degradation, unsustainable population growth, and destabilizing poverty beyond U.S. borders. He said that these issues threaten America's continued prosperity and its security, and that countries suffering from persistent poverty and worsening environmental conditions are not only poor markets for U.S. exports, but also likely victims of conflicts and crises that can only be resolved by costly American intervention. Thus, Secretary Christopher said, "the Clinton Administration is dedicated to restoring America's leadership role on sustainable development—an approach that recognizes the links between economic, social, and environmental progress. . . . Supporting the developing world's efforts to promote economic growth and alleviate chronic conditions of poverty serves America's interests."[4] The previous year Christopher asked Congress for foreign assistance funding to support "child survival, poverty lending, and micro-enterprise programs to help the poorest of the poor acquire sufficient food, shelter, and capital to become productive and healthy members of society and to provide for their children.   . . . Humanitarian assistance programs will always be part of our foreign policy because they project the values of the American people. They also reinforce our interest in sustainable development."[5]

These attitudes prevailed throughout the Clinton administration's tenure. Reflecting this, in July 1999 Secretary of State Madeleine Albright said that Americans "cannot be secure if the air we breathe, the food we grow, and the water we drink are at risk because the global environment is in danger. . . . The United States has the world's largest economy [and] the best environmental technology. And our society is by far the largest emitter of the gases that cause global climate change. So we have both the capacity and the obligation to lead."[6] She repeated these comments a year later in a speech for "Earth Day."[7] In September 1999, President Clinton said that Americans "have a big responsibility because America produces more greenhouse gases than any other country in the world."[8] These statements were reinforced by officials at various levels in the foreign policy bureaucracy.

While they are usually couched in terms of U.S. national interests, these statements nevertheless reflect growing concern about environmental changes and a nascent acceptance of *international equity*—a fair and just distribution among countries of benefits, burdens and decision-making authority associated with international relations—as one of the objectives of U.S. global environmental policy. This book argues that this acceptance of international equity objectives, albeit limited, is unprecedented in U.S. foreign policy, explains the reasons behind it, examines why the United States has failed to accept international environmental equity more robustly, and speculates on its future implications for U.S. interests and world politics.

In contrast to the Clinton administration's posture, the U.S. government under George Bush (and indeed Ronald Reagan) was extremely skeptical of the value of the whole United Nations Conference on Environment and Development endeavor, and opposed many of the equity provisions adopted by the conference or attempted to dilute them.[9] Yet even the Bush administration had agreed to provisions for international equity in the London amendments to the Montreal Protocol on Substances that Deplete the Ozone Layer, and by the time Bush attended the 1992 Earth Summit his administration's opposition to international equity considerations had softened substantially.

The U.S. acceptance in the 1990s of international equity as a goal of global environmental policy, albeit quite modest, is unusual by historical standards. The U.S. government has traditionally sought to deny responsibility for international inequities, especially insofar as they relate to financial commitments by the United States, and the U.S. government has been especially unwilling to seriously consider the demands of developing countries for more equitable treatment in international economic relations. Yet, in conjunction with increasingly well-understood and salient changes to the global environment, the U.S. government has softened and occasionally reversed its traditional opposition to matters of international equity.

The United Nations Conference on Environment and Development and the agreements coming from it, which are the main objects of this analysis of U.S. policy, were unprecedented events in international relations. Previous efforts to advance international equity norms in the environmental policy field, such as in the United Nations Conference on the Human Environment (UNCHE), the United Nations Conference on the Law of the Sea (UNCLOS), as well as more general calls for a New International Economic Order (NIEO), had little success. In contrast, the UNCED agreements and conventions signed at the June 1992 Earth Summit held in Rio de Janeiro elevated norms of international equity to prominence in the environmental issue area. What is more, it is possible that the provisions for international equity that were included in the UNCED agreements—if taken more seriously in coming decades—could signal a substantial shift in the conduct of international relations generally, not only in the environmental field.

Questions Addressed in this Book

This book looks at U.S. international environmental policy in the context of the UNCED process, including: (1) the negotiations leading to the 1989 UN General Assembly Resolution 44/228 establishing UNCED; (2) the preparatory committee (prepcom) negotiations dealing with the UNCED Declaration and Agenda 21; (3) the UNCED meeting held in June 1992 at Rio de Janeiro, Brazil (the "Earth Summit"), where associated international environmental agreements were signed; (4) the international deliberations regarding the world's forests leading to the UNCED statement on forest principles; (5) separate negotiations, including the intergovernmental negotiating committee (INC) meetings, for the Framework Convention on Climate Change and the Biodiversity Convention; and (6) subsequent international negotiations dealing with interpretation and implementation of these agreements.

In short, the book is concerned with the whole UNCED process, specifically from the late 1980s to the Earth Summit. It also examines with somewhat less focus the subsequent UNCED follow-on negotiations to show especially how United States policy has dealt with the equity provisions of agreements and conventions signed at the Rio convention. The UNCED process is ongoing and evolving; this book focuses on the formative stages of that process, especially as it relates to international equity and U.S. foreign policy, but it also looks a how and why U.S. policy has evolved since Rio. The specific primary questions this book seeks to answer include the following:

- To what extent have the U.S. and other governments accepted international equity as an objective of global environmental policy?
- What explains the U.S. government's acceptance of international equity as an objective of its policy in the global environmental field?
- Why did the U.S. government under President Bush not go further in accepting international equity as an objective of U.S. global environmental policy? Why did the Clinton administration go beyond the Bush administration in accepting international equity as an objective of global environmental policy? Why was the Clinton administration unable to *act* more robustly to promote international environmental equity?

Additional future-oriented questions addressed in this book include these:

- What are the practical policy implications and normative implications of the U.S. government's acceptance of international equity as an objective of its global environmental policy? That is, what effect might the partial U.S. embrace of international equity as an objective of its global environmental policy have on future U.S. definitions of its interests? How might this affect American influence in international environmental politics and international affairs generally?
- Should considerations of international environmental equity figure more prominently in American policymakers' calculations of U.S. national interests and global power?

As a prelude to answering these questions, the next chapter discusses the concept of international equity, which is defined in this book as *a fair and just distribution among countries of benefits, burdens, and decision making authority*. This concept has started to permeate global environmental policy making—including influencing the foreign policy of the United States. In reality, as international environmental deliberations, including UNCED, have shown, various interpretations of equity will be important in the formulation and justification of international agreements depending on the specific environmental issue subject to deliberation. Rarely will considerations of international equity be of *definitive* influence in international environmental negotiations (although from time to time they can be), but they can be very important considerations nonetheless. They have been codified in various international environmental instruments, such as the Montreal Protocol, Agenda 21, the Framework Convention on Climate Change, and the Convention on Biological Diversity.

To be sure, the motivations of diplomats and their governments are rarely based largely on altruism or a desire to promote international equity. However, it is not essential that there be altruistic *motivations* for an outcome to be equitable. Even in their domestic policies some countries provide special benefits to people in order to garner electoral support or to stave off revolution, and not primarily out of altruistic motives. Regardless of the original reasons for decisions to redistribute resources based on need, even when such decisions are based on the *self-interests* of ruling groups, such redistributions are commonly regarded as social justice or equity policies. We should not hold international society to a higher standard in determining whether policies or actions qualify as equitable. Yet, some governments (e.g., the Nordic countries) and individuals in governments (Al Gore, perhaps[10]) *are* sometimes motivated by altruism. Such governments and individuals try to promote equity norms by building coalitions with others who have different, even cynical, motivations. This process can institutionalize norms of behavior and policy that can affect subsequent policy decisions. Furthermore, as Keohane has observed, "Moralists such as Woodrow Wilson and Jimmy Carter sometimes gain high office; indeed, their moralism may appeal to the electorate. Furthermore, even officials without strong moral principles have to defend their policies, and it is often convenient to do so in moral terms. This requirement may lead them, in order to avoid cognitive dissonance, to take on some of the beliefs that they profess. The act of piety may engender piety itself. . . ."[11]

Diplomats at the Earth Summit and other international environmental negotiations no doubt had different definitions of international equity in their briefing papers and in their heads. Scholars can illuminate these various meanings of equity without arbitrating among the conflicting definitions. We ought to be reminded that international agreements, including those resulting from UNCED, are often vague in their definitions and statements, reflecting the differing values and interests of the signatories. Hence, if the reader is frustrated by a lack of clarity in defining "international equity," it is useful to bear in mind that the diplomats and bureaucrats have been equally frustrated.

Practical Significance of this Book

There are three interrelated practical justifications for undertaking a study of this kind: (1) normative ideas influence the foreign policies of governments; (2) there is an established and ever increasing need for the North to engage in environmental bargaining with the South, which can be facilitated by a sensitivity to the South's concern for equitable arrangements; and (3) how the United States adapts to the new imperatives of environmental diplomacy is likely to have an enormous impact on Earth's ecosystems and on prospects for human health and well-being in years to come, not only in the United States but all around the world.

Normative Ideas Influence International Relations and Foreign Policy

Ideas like international equity can become rooted and take on a life of their own in international relations, possibly becoming important determinants of—or at least constraints on—state behavior. As in domestic society, international norms may become more influential as a result of power bargaining, but once they are brought about they can become independent forces. Institutions like the environmental regimes emanating from UNCED can be conducive to this process. "Indeed," as Judith Goldstein and Robert Keohane argue, "one consequence of international institutions is that they provide settings in which governments must provide reasons (whether genuine or not) for their positions. The existence of international institutions gives states greater incentives to make their policies more consistent with one another and with prevailing norms, so that they can be more successfully defended in international forums."[12] Goldstein and Keohane see ideas as playing a role similar to that described by Max Weber. According to Weber: "Not ideas, but material and ideal interests, directly govern men's conduct. Yet very frequently the 'world images' that have been created by ideas have, like switchmen, determined the tracks along which action has been pushed by the dynamic of interest."[13] Ideas help to bring some order to politics and may be influential enough to shape agendas, thereby having a very significant impact on the course of events. Ideas can put "blinders" on people, limiting the number of policy options they have to choose from.[14]

Goldstein and Keohane describe three causal pathways whereby ideas find their way into foreign policy: (1) ideas serve as road maps; (2) ideas contribute to outcomes in the absence of unique equilibria; and (3) ideas embedded in institutions specify policy in the absence of innovation.[15] These pathways suggest ways in which considerations of international equity may influence the creation and operation of international environmental institutions.

While the first pathway does not account for how certain ideas become salient, it does limit choice of policy by "excluding other interpretations of reality or at least suggests that such interpretations are not worthy of sustained exploration."[16] Although exceptions abound, it seems that states face difficulty if they attempt to negotiate agreements that are blatantly unfair to other participants. International environmental agreements such as those seeking to limit global climate change can be successfully negotiated only with the participation of many developing countries (e.g., China, India, Brazil) whose negotiating power has increased in proportion to the importance of their participation. The developing countries must think they are getting a fair deal if they are to sign on to these agreements and undertake genuine fulfillment of them. This fact has contributed to considerations of equity becoming an important part of nascent and established international environmental institutions, including the amended Montreal Protocol and the agreements and conventions signed at the Earth Summit. Equity has become one of the guidelines for negotiators. To be sure, each diplomat at international forums will try to promote his own state's interests, but he will do so in the context of treating other states—including those weaker than his own—fairly. Equity does not *dictate* the progress and outcomes of international environmental negotiations, but it does seem to act as an important *restraint* or *guideline*, at the very least turning states away from blatantly unfair choices.

Normative ideas may be especially useful as guidelines in situations of uncertainty. International bargaining directed toward international environmental change are characterized by uncertainty. Especially when the science is murky (as it has been on the issue of climate change, for example), actors have some difficulty identifying their long-term interests. They seek guidelines within which their various conceptions of the future can operate. Goldstein and Keohane contend that "Ideas serve the purpose of guiding behavior under conditions of uncertainty by stipulating causal patterns or by providing compelling ethical or moral motivations for action. . . . Causal ideas respond directly to uncertainty by reducing it, whereas principled ideas enable people to behave decisively despite causal uncertainty. Principled ideas can shift the focus of attention to moral issues and away from purely instrumental ones focused on material interests and power."[17] Equity is one such compelling idea that can serve to move deliberations away from narrow discussions of self-interest.

The second pathway described by Goldstein and Keohane suggests that ideas act as "focal points" or "glue." By their account, "change can occur when perceptions of new benefits to be realized from cooperation give coordination problems new salience and lead actors to search for ideas that will enable them to cooperate."[18] The developed states want and need the participation of several developing states to make environmental institutions (such as the Climate Change Convention) work. Developing states want financial aid to at least cover their incremental costs of participation. Equity provides a focal point around which their expectations can converge. The related notions of "environment and development" and "sustainable development" emphasized at the Earth Summit can serve as a "conceptual bridge" between the environmental goals of the developed countries and the economic development objectives of the developing world.[19]

Finally, and perhaps most important for the promotion of international equity in the long term, is what happens once a distributional principle becomes institutionalized. Once an idea becomes a component part of institutional rules and norms its initial basis (whether self-interest or temporary altruism) may no longer be salient. Such principles often remain to structure subsequent political debates and institutional deliberations, and thus continue to affect the evaluation of policy choices by those who use the institutions. As put by Goldstein and Keohane, "the interests that promoted some statute may fade over time while the ideas encased in that statute nevertheless continue to influence politics. Thus at a later time, these institutionalized ideas continue to exert an effect: it is no longer possible to understand policy outcomes on the basis of contemporary configurations of interest and power alone."[20]

Thus, even if equity components in the Montreal Protocol, the UNCED agreements and conventions, or other international environmental instruments were a disingenuous ploy by representatives of developed states to garner signatures from developing countries, and even if incorporation of the equity provisions were mostly rhetorical wrapping for agreements based essentially on self-interest, in the long-run such equity provisions may very well prove to be influential. They may yet be important in the workings of the implementation regimes and may bring substantial economic benefits to people in many developing countries, while simultaneously limiting global environmental changes that may adversely affect all people.

What is being suggested here is that where ideas like international equity can play a role in the formation of international environmental institutions, even if other factors (such as power and material self-interest) are important, normative ideas are likely to become embedded in these institutions—and in international politics generally—and thus continue to affect policy choices. They may even increase in salience over time, to become relatively more determinative of policy outcomes.

The Need to Bargain with the South

Global environmental issues are becoming increasingly salient in international relations. For example, climate change, caused by the introduction of greenhouse gases (GHGs) into the atmosphere, is perhaps the greatest long-term threat to the global environment on which humankind depends for its prosperity and survival. In the coming decades, climate change may result in dramatic changes in sea level, ocean currents, and weather patterns, with consequences ranging from more frequent and severe floods and droughts to the spread of pests and the submergence of some island countries.[21] Indeed, climate change poses such potentially unprecedented challenges to the international community that we can expect the negotiations on climate change to last well into and perhaps throughout this century, much as the General Agreement on Tariffs and Trade (GATT) and subsequent World Trade Organization (WTO) talks have been ongoing for over half a century. Several developed country diplomats involved in the United Nations climate change negotiations conceded that equity considerations are a crucial component of successful negotiations and agreements meant to limit climate change.[22]

Consequences of transboundary environmental pollution, such as stratospheric ozone depletion, ocean pollution, and climate change, can be limited or prevented only if both economically developed and the large developing countries reduce their polluting emissions. Unilateral efforts by the developed industrialized countries, while essential as a first step, will be overwhelmed as the large developing countries use more energy and produce more environmental pollutants. If China burns its vast coal reserves and Brazil cuts its expansive rain forests, greenhouse gas levels will increase beyond the potential control of the industrialized countries. The developed countries must do much more to reduce their own emissions of greenhouse gases. But the developing countries must also be persuaded that they should forgo the energy-intensive industrialization enjoyed by the developed countries, instead developing in a manner that does not rely as heavily on fossil fuels. Such persuasion will require substantial concessions on the part of the developed countries, involving major transfers of funds and new, more environmentally benign technologies. In other words, the extent to which there is sustainable development in the developing world is a *global* concern that will require more serious attention from the South— and this will be more true as the South grows and adopts consumption patterns analogous to those in the profligate North.

Maldistribution of social, economic, political and environmental resources is often synonymous with unsustainable development. The poor are concerned about fulfilling their basic needs and, once that is accomplished, raising their standards of living. They are unlikely to be concerned with environmental changes whose adverse effects will be experienced or suffered in the relatively distant future, especially when those problems are largely caused by (and concern) the wealthy people of the world who the poor often blame for much of their suffering. The people of the developing world believe that it is unfair for the citizens of the developed countries to ask the poor to forgo development so that the North can continue to consume as it has so far. Only if the poor are treated fairly by the rich will they genuinely join in efforts to protect the global environment. They cannot be expected to participate in international environmental agreements if such agreements are perceived as being unfair. As Oran Young points out:

Those who believe that they have been treated fairly and that their core demands have been addressed will voluntarily endeavor to make regimes work. Those who lack any sense of ownership regarding the arrangements because they have been pressured into pro forma participation, on the other hand, can be counted on to drag their feet in fulfilling the requirements of governance systems. It follows that even great powers have a stake in the development of international institutions that meet reasonable standards of equity.[23]

Environmental changes create a situation in which considerations of equity at the international level have greater importance than they might have without that environmental change. In contrast to other issues, questions of whether and how much the developed countries should aid the developing countries are unavoidable in the global environmental policy arena. What is more, with the end of the cold war there are potentially new sources of aid (e.g., the illusive, almost forgotten "peace dividend") and, paradoxically, new justifications for discontinuing aid (it is no longer essential for opposing superpowers to garner friends through international aid). While the *old* North-South debate is still salient, the world has changed enough to require a *new* debate about aid from developed to developing countries.[24]

Climate change demonstrates how environmental issues can foster the salience of equity in international politics. The continuing climate change negotiations are substantially different than past experiences with the UNCHE, the NIEO, and the Law of the Sea. As Shue has argued:

> A political decision to adopt a global ceiling on GHG emissions has implications for equity that are far more radical than has so far been recognized. A serious decision to deal with the natural limit on the planet's capacity to dispose of GHG emissions by imposing a political limit on the emissions produced by humans totally transforms the international situation. The reason is simple: imposition of an emissions ceiling makes emissions, as the economists like to say, zero-sum. For equity this change has powerful implications.[25]

Because equity fundamentally requires that one do no harm, according to Shue, "The adoption of a ceiling on total emissions moves the consumption of more than one's share of allowable emissions into a new category of equity, the category of rock-bottom prohibited wrong."[26]

Furthermore, the uncertainty associated with international environmental issues (noted above) can have important influences on international bargaining. It can be an instrument for the creation of international environmental institutions that are more equitable. Young has suggested that:

> Uncertainty may also play a constructive role in making it difficult for participants in institutional bargaining regarding climate change to make confident predictions about the distributive consequences of alternative institutional arrangements under consideration for inclusion in a climate regime. The resultant veil of uncertainty has the effect of increasing interest in the formation of arrangements that can be justified on grounds that they are fair in procedural terms, whatever substantive outcomes they produce. Coupled with the operation of the consensus rule characteristic of institutional bargaining, this has led some analysts to argue that effective international agreement to limit [greenhouse gas emissions] will not be undertaken unless the agreement is seen by the participants as fair.[27]

Importantly, international environmental institutions help protect the global environment and may help prevent violent conflict that can result from environmental degradation.[28] We can define international institutions (often referred to as "regimes") as recurring sets of principles, norms, rules, and decision making procedures that act as guidelines for how states behave toward one another.[29] International institutions usually coincide with formal international treaties or "soft law" agreements, and often have accompanying organizations. Several international institutions have been created that attempt to deal with adverse global environmental changes. One of the best known and most successful is the 1987 Montreal Protocol on Substances that Deplete the Ozone Layer (as amended).[30] The Protocol has led to major reductions in gases that harm the stratospheric ozone layer. Other international environmental institutions have contributed to the protection of the global environment. As the distribution of power in the international system becomes more diffuse in many issue areas, the effectiveness of many international institutions (notably those that had limited effectiveness during the U.S.'s postwar hegemony, including those addressing environmental concerns) may increase. We may be entering a period in which international institutions generally will be more important to the successful conduct of interstate relations.[31]

However, there is no certainty that institutions will form, and once formed there is no certainty that they will be successful. It is therefore important and worthwhile to explore all factors that may increase the likelihood of institutional formation and effectiveness. We already know that equity considerations are important for the creation and effectiveness of international environmental institutions.[32] Countries are more likely to participate in international environmental institutions if associated arrangements are seen as fair and just. International equity considerations are therefore increasingly a prominent component of international environmental institutions, most notably the nascent institutions created during UNCED. Hence, a focus on equity considerations is a useful undertaking, even apart from the many important ethical considerations. As Oran Young tells it, "The study of governance systems in international society cannot prosper in the absence of a better understanding of the determinants of effectiveness. Those responsible for designing governance systems to cope with growing threats to the earth's habitability demand knowledge that they can use to devise regimes that will prove effective."[33] The upshot is that if we are to protect the global environment we need to better understand the processes by which international environmental institutions can be made to appear equitable in the eyes of important parties.

A research project directed by Oran Young and Gail Osherenko brought together scholars from several countries to examine the following question: What are the determinants of success or failure in efforts to form regimes dealing with specific environmental and resource issues?[34] To answer this question they examined several institutional arrangements geared toward protecting polar ecosystems. They grouped their hypotheses and ultimate explanatory arguments into four categories: power-based hypotheses, interest-based hypotheses, knowledge-based hypotheses, and contextual arguments.[35] Of the twelve interest-based hypotheses examined across five case studies, with the exception of individual leadership no explanation was found to be more important than equity in explaining the successful formation of international environmental regimes.[36] Young and Osherenko found that "institutional bargaining cannot succeed unless it produces an outcome that participants can accept as equitable, even when the adoption of equitable formulas requires some sacrifice in efficiency."[37]

States usually do not comply with international environmental standards (or most other international standards) because they are forced to do so. Instead, they comply with more subtle pressures coming from "a combination of binding international law and public exposure of noncompliance (often by less inhibited nongovernmental organizations [NGOs]), normative persuasion, scientific argument, technical assistance, and investment."[38] Institutions can facilitate cooperation and compliance with international environmental agreements by linking the agendas of those institutions with issues of greater concern to governments. Material incentives in international institutions, such as financial aid and technology transfer to developing countries and the new democracies of Eastern Europe, and the trade sanctions found in the Montreal Protocol, are examples of how such direct linkages can be made. Such factors can be entirely consistent with equity considerations in international environmental agreements.[39]

In addition to fostering institutional formation, international equity is an important contributor to the effectiveness of international environmental institutions.[40] Effectiveness can be defined as changes in state behavior that would not occur without the institution, and which help limit or prevent damage to the natural environment.[41] In one of his own case studies analyzing management of Arctic shared natural resources, Young found that effectiveness is enhanced if negotiators devise arrangements that all actors can accept as being equitable and legitimate in the long-term.[42] While such legitimacy may be unnecessary for narrow, short-term arrangements, it is critical in situations where continuing conformity to agreements is necessary over an extended period, which is the case with most global environmental problems.

The progression of equity considerations through two decades of international environmental negotiations demonstrates that countries are at least beginning to recognize the importance of considering equity. At the 1972 UN Conference on the Human Environment, international equity considerations played a minor role. As with the NIEO, demands by developing countries for technology transfers, new funding and adjustments to the world economy were largely ignored by the developed countries. International equity considerations were more prominent in the Law of the Sea Convention, and the 1987 Montreal Protocol included some provisions for equity. But it was the amendments to the Montreal Protocol agreed at London in 1990 that were permeated with considerations of international equity. In contrast to 1987 when parties to the Montreal Protocol sought only a reduction in ozone-destroying chemicals, the 1990 meeting sought to eliminate them altogether. Parties realized that this objective could only be achieved with the participation of the developing countries. Hence, international equity considerations became a central, almost predominant, component of the ozone treaty. It was against this backdrop that UNCED negotiations were conducted.

How can we explain this growing prominence of international equity in the environmental field? Traditional evaluations of state power based on economic and military resources are not sufficient. If the distribution of power is the best explanation for the shape of international environmental institutions,

> how can we explain cases such as the bargaining over the deep seabed mining provisions of the 1982 Convention on the Law of the Sea or the interactive process eventuating in the collapse of the 1988 Convention on the Regulation of Antarctic Mineral Resource Activities, in which acknowledged great powers—the United States in the law of the sea case

and the United States and Great Britain in the Antarctic minerals case—are unable to prevail on others to accept their preferred arrangements? And what are we to make of cases such as the negotiations that produced the 1990 London Amendments to the 1987 Montreal Protocol on Substances that Deplete the Ozone Layer or the 1991 Environment Protocol to the Antarctic Treaty, in which others are able to pressure a great power—the United States again—into accepting provisions it initially opposes?[43]

Sometimes otherwise powerful countries are hamstrung by internal debates over specific foreign policies, thereby limiting their power to influence the policies of other countries and the shape of international institutions. During UNCED negotiations on the Biodiversity and Climate Chang Conventions, the United States experienced internal dissension over appropriate policy, manifested most notably by embarrassing press disclosures of significant differences between the head of the Environmental Protection Agency (EPA) and White House personnel.[44] The United States failed to demonstrate strong leadership and the conventions were signed despite U.S. objections. Subsequent negotiations have been permeated by discussions of international equity, due in part to the Bush administration's failure to lead in another direction, which has helped set (or at least not prevent) a modest precedent for greater consideration of equity in international bargaining on global environmental issues.

As Young points out, theoretical perspectives that emphasize the role of state power may be inadequate:

> Those who emphasize the role of power in international affairs are apt to dismiss considerations of equity as normative concerns that have little bearing on the course of events. In institutional bargaining, however, there are good reasons for participants to take a genuine interest in matters of equity, even if they possess abundant sources of structural power. Partly, this is because institutional bargaining at the international level—unlike legislative bargaining in most domestic arenas—proceeds under a consensus rule. Such bargaining can succeed only when it yields contractual formulas acceptable to all the relevant parties or coalitions of parties. Of course, those with structural power may be able to buy acquiescence from others by providing them with compensation. This is exactly what happened in the case of ozone and what must happen if continuing climate change negotiations are to produce an effective governance system. But such arrangements already constitute a move in the direction of equity in the sense that they involve a departure from the image of great powers simply calling the shots without any concern for

the interests of others.[45]

Influential members of the international community, rather than relying on their traditional power resources (e.g., military strength), may be more successful if they use their capacities to provide economic and technological incentives to developing countries whose participation is essential to efforts to protect the global environment. It is seldom possible to force developing countries to participate. Efforts to address transboundary air pollution, deforestation, and other environmental problems have shown that powerful countries can obtain the greatest level of compliance by helping poor countries overcome the technological and financial hurdles associated with implementation of international environmental agreements. Threats or sanctions geared toward forcing compliance may have some efficacy, but will not be as reliable as capacity building and financial aid.[46] Thus equity is a key component of efforts to bring the South on board international environmental agreements.

The Crucial Role of the United States

The rest of the world expects the United States to be a leader on global environment and development.[47] According to one observer of the UNCED process, the United States "is looked at by the rest of the world as the logical leader on these kinds of issues, because of our size and weight in the world economy, because we still have the world's greatest scientific and technological capability, because we still have the world's greatest diplomatic influence of any single country, because we have the longest experience in managing the environment. They do look to us for leadership. . . ."[48] As the most powerful country in the world by most measures, including military, political and economic, U.S. policies and behavior have an inordinate impact on other countries in most issue areas, especially the environmental area.

It is not possible to effectively address the most pressing global environmental problems without U.S. participation because the United States is one of the largest polluters—per capita and in aggregate—of global environmental commons. The average American uses many times the amount of energy, and produces many times as much pollution and waste, than do people in most other countries. Furthermore, U.S. businesses are busy producing goods for the American and world markets. In so doing they produce prodigious amounts of waste and pollution. They are still often inefficient and dirty relative to firms in other highly industrialized countries, but what matters the most is the utter scale of their economic output. Even though the United States has been quite successful in reducing pollution *within* its borders, overall it produces more *global* pollution than any other country. Thus, if it continues to pollute, the global environment will continue to suffer greatly.

Moreover, the United States has weighty influence in international efforts to address global environment and development issues. As the world's largest economy, the United States is the world's largest foreign aid donor (although on a per capita basis it falls behind other developed countries and even in aggregate it was edged out by Japan). What is more, the United States provides the guarantees for massive amounts of funds provided by the world's international financial institutions, and as the largest donor to many of these institutions it generally has extraordinary influence in decisions about how they (notably the World Bank and International Monetary Fund) administer those funds. The United States is also the largest financial supporter of basic UN operations (when it pays up, which it does for the most part after much complaining). Despite the grudging nature of this assistance, it is essential to UN efforts in most areas, including environmental protection and economic development. Moreover, a huge portion of global private investment comes from U.S.-based or U.S.-affiliated corporations, and how these multinationals invest and operate can have profound effects on environment and development where they operate. In short, the United States has the cash that is needed to help promote *environmentally sustainable* development.

In addition, the United States has much of the knowledge, expertise, and technologies needed to move the world into a "greener" future. Problems of global environmental change are intimately tied up with economic development. Responsible governments everywhere will try to improve the well-being of their citizens. This is an unavoidable (and generally laudable) goal. If this economic development is not to cause severe and sometimes devastating environmental harm, however, it must be done in a way that minimizes pollution and use of natural resources. This requires the deployment of energy-efficient and "environmentally-friendly" technologies. As one of the world's technological powers, and a source of much of the innovation in this area, the United States has a central role to play. It has, or will develop, many of the technologies that will make economies more efficient and the world less polluting. The U.S. government and industry must of course deploy these technologies at home, but they also can share them with the rest of the world, possibly allowing developing countries to leapfrog some of the damaging industrialization practiced in developed countries. Especially if it does so on concessional terms, this has the potential to avoid untold harm to the natural environment and to improve the lives of vast numbers of people.

We should also bear in mind that the United States normally wields exceptional diplomatic and political influence in the world. By setting an example in its use of energy, levels of pollution, and assistance to the developing world, it can lead other wealthy countries into cooperative endeavors to protect the global environment and promote environmentally sustainable development. There have been occasions when it has done this to a great effect, showing the potential of its leadership.[49] Having said this, while the United States is essential to successful efforts to address global environmental problems, it cannot easily "get its way" in global environmental politics. Indeed, other countries—clearly weaker in all traditional measures of international power—have been able to prevent it from doing just that. This was evident during the UNCED process. While the U.S. government did have its usual inordinate influence in shaping the outcomes of negotiations, it failed to achieve many of its goals. It failed in its attempt to get the Rio Declaration on Environment and Development called the "Earth Charter" (which would have lessened the visibility of the developing countries developmental goals at the conference); it failed miserably in its efforts to negotiate a forest convention, instead agreeing to a feeble statement on forest principles; it failed to prevent the signing of the Biodiversity Convention, and it failed in most cases to limit the setting of precedents in the UNCED agreements that might promote international equity and thereby increase U.S. obligations to other countries (among other failures).

We should not overstate these failures, but they do show that the United States could not easily push other countries around, including many of the traditionally weak developing countries. It needed to consider other countries demands for fair consideration of their own interests and priorities. (The inability of the United States to shape the UNCED process to its liking arguably challenges the assumptions of "realist" thinkers about the role of power, at least in its traditional forms, in international relations.) What this shows is the importance of U.S. *leadership* and *cooperation* with other countries, and it suggests that U.S. efforts to push its weight around, at least in this issue area, can be ineffectual and even counterproductive.

Conclusion

The U.S. government has come to see environmental changes as important subjects of concern, and it has started to realize that they must be higher on the global political agenda. Furthermore, it has come to join an emerging international consensus that supports concrete efforts to incorporate equity into global environmental politics. These changes in official attitudes are extremely important, not only because global environmental problems increasingly threaten the world but also because U.S. pollution, economic and technological resources, and political power in the world bear directly on these problems. However, the U.S. government has come only so far. While the Clinton administration did more than its predecessors to make equity a part of U.S. global environmental policy, it could have done much more. After looking at the evolution of international equity in global environmental politics, the final chapters of this book examine and explain this limited change in U.S. policy.

Notes

1   Timothy Wirth, "World Conference on Human Rights," press briefing, Washington, DC, 2 June 1993, *U.S. Department of State Dispatch* 4, 23 (7 June 1993).
2   Albert Gore, "U.S. Support for Global Commitment to Sustainable Development," address to the Commission on Sustainable Development, United Nations, New York City, 14 June 1993, *U.S. Department of State Dispatch* 4, 24 (14 June 1993).
3   William J. Clinton, "Advancing a Vision of Sustainable Development," address to the National Academy of Sciences, Washington, DC, 29 June 1994, *U.S. Department of State Dispatch* 5, 29 (18 July 1994).
4   Warren Christopher, "Overview of 1995 Foreign Policy Agenda and the Clinton Administration's Proposed Budget," statement before the Senate Foreign Relations Committee, Washington, DC, 14 February 1995, *U.S. Department of State Dispatch* 6, 8 (20 February 1995).
5   Warren Christopher, "Foreign Affairs Budget that Promotes U.S. Interests," statement before the Subcommittee on Foreign Operations of the Senate Appropriations Committee, Washington, DC, 2 March 1994, *U.S. Department of State Dispatch* 5, 11 (14 March 1994).
6   Madeleine K. Albright, "Secretary of State Albright's Remarks to the National Association for the Advancement of Colored People (NAACP) in New York, 13 July 1999", *Environmental Change and Security Project Report* 6 (Summer 2000), p. 122.
7   Madeleine K. Albright, "An Alliance for Global Water Security in the 21st Century," Earth Day speech, Washington, D.C., as released by the Office of the Spokesman, U.S. Department of State, 10 April 2000, <http://secretary.state.gov/www/statements/2000/ 000410.html>.
8   William J. Clinton, "President Clinton's Remarks to the People of New Zealand, Antarctic Centre, Christchurch, New Zealand, 19 September 1999," *Environmental Change and Security Project Report* 6 (Summer 2000), p. 119.
9   This book went to press as the administration of Bush's son, George W. Bush, was taking office. References are to the senior Bush and his administration.
10 At least that is the perception that one can get from reading his book, *Earth in the Balance* (New York: Houghton Mifflin, 1992). His statements as vice president, if not all the Clinton administration's policies, seem to bolster this view.

11   Robert Keohane, *After Hegemony* (Princeton: Princeton University Press, 1984), p. 127.

12   Judith Goldstein and Robert Keohane, *Ideas and Foreign Policy* (Ithaca: Cornell University Press, 1993), p. 24.

13   Max Weber, "The Social Psychology of the World Religions," in H.H. Gerth and C. Wright Mills (eds.), *From Max Weber· Essays in Sociology* (New York: Oxford University Press, 1946), p. 280.

14   Goldstein and Keohane, p. 12.

15   Albert Yee describes three general ways by which ideas can affect policies through institutions: (1) the bureaucratic power of "epistemic communities"; (2) ideas "encased" within institutions and policy-making bearers of ideas; and (3) institutional constraints on the access, flow, and impact of ideas within the policy-making process. Albert S. Yee, "The Causal Effects of Ideas on Policies," *International Organization* 50, 1 (1996), pp. 69-108. Epistemic communities are networks of knowledge-based experts "with recognized expertise and competence in a particular domain and an authoritative claim to policy-relevant knowledge within that domain or issue-area." They help states identify their interests, frame the issues for collective debate, propose specific policies, and identify salient points for negotiation. Peter M. Haas, "Introduction: Epistemic Communities and International Policy Coordination," *International Organization* 46, 1 (1992), p. 3.

16   Goldstein and Keohane, p. 12.

17   Ibid., p. 16-17.

18   Ibid., p. 25.

19   Cf. Marvin Soroos, "From Stockholm to Rio: The Evolution of Global Environmental Governance," in Norman J. Vig and Michael E. Kraft, eds., *Environmental Policy in the 1990s* (Washington: CQ Press, 1994), pp. 299-321.

20   Goldstein and Keohane, p. 21.

21   Intergovernmental Panel on Climate Change (IPCC), "IPCC Working Group I 1995 Summary for Policymakers," fifth Working Group I session, Madrid, 27-29 November 1995; J.J. Houghton et al., eds., *Climate Change 1995: The Science of Climate Change* (New York: Cambridge University Press, 1996); R.T. Watson et al., eds., *Climate Change 1995: Impacts, Adaptations and Mitigation of Climate Change* (New York: Cambridge University Press, 1996).

22   I base this on, among other things, discussions with several delegates and their deputies involved in the United Nations Intergovernmental Negotiating Committee (INC) on Climate Change, New York, 9-10 February 1995, including, Rafe Pomerance, U.S. Deputy Assistant Secretary of State for Environment and Development, interview by author, New York, 10 February 1995.

23   Oran R. Young, *International Governance* (Ithaca: Cornell University Press, 1994), p. 134.

24   The "new" North-South debate is characterized by, for example, more reasonable demands from the poor countries and a clear emphasis on using official development assistance in more environmentally friendly ways or for projects designed specifically to protect the local, regional, and/or global environment. For respected characterizations of the "old" North-South debate, see Willy Brandt et al, eds., *North-South: A Programme for Survival* (Cambridge, MA: MIT Press, 1980) and *Common Crisis North-South: Cooperation for World Recovery* (Cambridge, MA: MIT Press, 1983).

25 Henry Shue, "Equity in an International Agreement on Climate Change," paper prepared for the IPCC Working Group III, draft of 15 July 1995, mimeo, pp. 2-3.

26 Ibid., p. 7.

27 Young, *International Governance*, p. 43-44. Young notes that Brennan and Buchanan observe in a discussion directed toward municipal institutions that "to the extent that a person faced with constitutional choice remains uncertain as to what his position will be under separate choice options, he will tend to agree on arrangements that might be called 'fair' in the sense that patterns of outcomes generated under such arrangements will be broadly acceptable, regardless of where the participant might be allocated in such outcomes." Ibid., pp. 101-102.

28 On institutional effectiveness, see, for example, Peter M. Haas, Robert O. Keohane, and Marc A. Levy, eds., *Institutions for the Earth: Sources of International Environmental Protection* (Cambridge: MIT Press, 1993); and Oran Young, "The Effectiveness of International Institutions: Hard Cases and Critical Variables," in James N. Rosenau and Ernst-Otto Czempiel, eds., *Governance Without Government: Order and Change in World Politics* (Cambridge, England: Cambridge University Press, 1992). On environment and conflict, see, for example, Thomas Homer-Dixon, "On the Threshold: Environmental Changes as Causes of Acute Conflict," *International Security* 16, 2 (1991), pp. 76-116; Norman Myers, *Ultimate Security: The Environmental Basis of Political Stability* (New York: W.W. Norton, 1993); and Woodrow Wilson Center, *Environmental Change and Security Project Report*, 1 (1995), special issue dedicated to "Environment and Security Debates."

29 Stephen D. Krasner, ed., *International Regimes* (Ithaca: Cornell University Press, 1983). As John Mearsheimer, "The False Promise of International Institutions," *International Security* 19, 3 (1995), pp. 3-49, points out, there is no common definition of international institutions. For discussions of international institutions as they relate to international environmental issues, see Haas, Keohane and Levy; Oran R. Young and Gail Osherenko, *Polar Politics: Creating International Environmental Institutions* (Ithaca: Cornell University Press, 1993); Young, *International Governance*; Seyom Brown et al., *Regimes for the Ocean, Outer Space, and Weather* (Washington: Brookings Institution, 1977); and Ken Conca et al., eds., *Green Planet Blues: Environmental Politics from Stockholm to Rio* (Boulder: Westview Press, 1995), pp. 165-203.

30 Richard Benedick, *Ozone Diplomacy* (Cambridge: Harvard University Press, 1998).

31 Young, "Effectiveness of International Institutions," p. 187. See generally Oran R. Young, *Governance in World Affairs* (Ithaca: Cornell University Press, 1999).

32 See, for example, Young and Osherenko; and Young, *International Governance*.

33 Young, *International Governance*, p. 160.

34 Young and Osherenko. Rather than examine the factors leading to both regime formation and effectiveness, their project focused on the former.

35 Ibid., pp. 8-21, 263-66.

36 Of the twelve interest-based hypotheses, "individual leadership" was confirmed by all five cases and "salient solutions" and "integrative bargaining" were confirmed as often as the "equity" hypothesis (each was confirmed by four of the cases, with mixed results on one other case). Only the "values/ideas matter" hypothesis in the knowledge-based category proved to be nearly as powerful (confirmed by four of the case studies). See Young and Osherenko, chapter seven.

37 Ibid., p. 235.

38 Haas, Keohane and Levy, p. 17.

39   Ibid., p. 400-401.
40   See, for example, Richard L. Hembra, *International Environment: Options for Strengthening Environmental Agreements* (Washington, DC: General Accounting Office, July 1992).
41   Success may also depend, according to some measures, on the degree to which poverty is reduced in the process of environmental protection.
42   Young, *International Governance*, p. 62.
43   Ibid., p. 117.
44   James Brooke, "U.S. Has a Starring Role at Rio Summit as Villain," *New York Times* (2 June 1992), p. A11. The EPA/White House brouhaha was but the tip of an iceberg. Several U.S. diplomats involved in shaping the UNCED agreements were sympathetic to many of the equity demands of developing countries. Interviews by author with former U.S. delegates to UNCED negotiations, November 1995 and January 1996, cited below, passim (see especially chapters 6-8).
45   Young, *International Governance*, p. 133.
46   Oran Young, "International Regime Initiation," *International Studies Notes* 19, 3 (1994), p. 45.
47   I summarize U.S. international environmental diplomacy in Paul G. Harris, "International Environmental Affairs and U.S. Foreign Policy," in Paul G. Harris, ed., *The Environment, International Relations, and U.S. Foreign Policy* (Washington, DC: Georgetown University Press, 2001), pp. 3-42.
48   U.S., Congress, Committee on Foreign Affairs, Subcommittee on Western Hemisphere Affairs, testimony of Gareth Porter, Director, International Program, Environmental and Energy Study Institute, *The United Nations Conference on Environment and Development*, 102nd Cong. (4 February 1992), 1993, p. 53.
49   See, for example, John Barkdull, "American Foreign Policy and the Ocean Environment: A Case of Executive Branch Dominance," in Paul G. Harris, ed., *The Environment, International Relations, and U.S. Foreign Policy* (Washington: Georgetown University Press, 2001).

2   Defining International Environmental Equity

The following chapters will show that international equity has become an important consideration in global environmental politics and in U.S. international environmental policy. They will also illuminate possible explanations for this process. This chapter endeavors to start defining the notion of international environmental equity.[1] By "equity" I mean "fairness" or social and distributive "justice." Philosophers will not like this lumping together of these terms. However, in the real world—including international environmental negotiations—they are routinely used interchangeably. I define "international environmental equity" as *a fair and just distribution among countries of benefits, burdens, and decision-making authority associated with international environmental relations.* This definition captures most of the various interpretations of equity (and related terms) used in international environmental deliberations and agreements.[2]

Global environmental change has a profound effect on interpretations of equity, justice and fairness. David Miller points out that for a state of affairs to be unjust it must result from the actions of persons, or at least be capable of being changed by human actions. He goes on to illustrate this point by example:

> Thus although we generally regard rain as burdensome and sunshine as beneficial, a state of affairs in which half of England is drenched by rain while the other half is bathed in sunshine cannot be discussed (except metaphorically) in terms of justice—unless we happen to believe that Divine intervention has caused this state of affairs, or that meteorologists could alter it.[3]

It is ironic indeed that such a discussion would be hardly metaphorical today, barely more than two decades after Miller's writing! To put it bluntly, today we *can* alter the weather due to our contributions to global warming and resulting climate change. In other words, industry and over-consumption on this side of the world causes foul weather on that side of the world.[4] Climate change (global warming) and international collaboration to deal with it and other environmental changes pose profound burdens and potential benefits for almost all countries, thus presenting us with important practical and ethical questions of international equity.[5] Hence, Miller's example of where equity does not apply is precisely where it ought to—and does—in the current and especially future contexts of global environmental change.[6]

Given current and anticipated adverse changes to the natural environment, can we find support for more serious considerations of—and action on—equity between the world's rich and poor countries? What properties can fit definitions of international equity in the context of environmental change? That is, when will we recognize a fair and just distribution among countries of benefits, burdens and decision-making authority associated with, in this case, international environmental relations? When can we apply the definition? The answer is, not surprisingly, "it depends." Several different and often conflicting meanings of equity are possible. Rather than try to justify any particular interpretation or principle of what is fair and equitable in the context of international environmental relations, this chapter endeavors to examine several interpretations. As the real world of international environmental politics—such as deliberations associated with UNCED and the resulting agreements—show, multiple, disparate, and often competing definitions of equity have been germane.

Indeed, it may be neither possible nor desirable to have only one definition of equity. To seek a single definition of equity may be, according to one observer, a "hopeless and pompous task."[7] When surveying the literature on social justice, Bernard Cullen notes that:

> The dominant impression is of something approaching philosophical pandemonium. When researchers in other disciplines . . . concerned with issues of justice and injustice approach the philosophers in search of a generally accepted definition and analysis of the object of their study, they are faced with a cacophony of discordant philosophical voices. Probably the most apt term to characterize the dozens of theories . . . , when taken together, is "incommensurability."[8]

Hedley Bull said that justice is a term that can be given only some kind of private or subjective definition.[9] When examining problems of equity, according to Stanley Hoffman, "even within the same ideological camp, one man's remedy is another man's poison. We find here the interaction of embattled systems of ethics, intractable political realities, and extraordinary scientific ignorance."[10] What constitutes equity even *within* national societies is subject to profound disagreement.[11] Such is also the case with more widespread applications of equity.

While different scholars have their preferences, which

interpretations are salient will depend on the specific issues being discussed, the actors involved and their relevant power vis-à-vis other actors in the issue area, and the bargaining environment in which deliberations take place. There is no single universal interpretation of what constitutes inequity in international relations (although there may be some specific interpretations, dealing with very specific issues, that approach near-universal acceptance[12]). To suggest that there is one universal principle of equity generally or one interpretation of international environmental equity specifically may be dangerous in that it might constrain debate. It might inhibit those actors who feel threatened or harmed by that interpretation from entering political deliberations over what is fair and equitable, thereby stifling cooperative efforts to address adverse environmental change and the poverty that frequently accompanies it. Hence, as the forthcoming discussion tries to show, while environmental changes are acting as stimuli for serious considerations of international equity, multiple interpretations of equity are considered by scholars and practitioners of international politics.

Miller is comfortable with a diversity of interpretations of equity: "Most people, I suggest, give some weight to each of [several] principles of justice, and decide how to act in concrete situations by weighing up various considerations and allowing the just action to emerge as a resultant."[13] Thompson points out that definitions of equity ought to take into account the prevailing values, allegiances, objectives and ideas about equity that people and the communities in which they live hold, but we should not assume that these values and allegiances are fixed or that they should be uncritically accepted.[14] To focus on only one interpretation of international environmental equity would be holding international relations to a higher standard than we hold even domestic politics. Competing definitions of what is fair and equitable are common in domestic society; witness debates in the United States over capital punishment, rights to own and carry weapons, entitlements to social welfare, and abortion. Indeed, there is a growing movement trying to define and promote environmental equity within domestic communities, including the United States.[15] Political confrontations over what ought to constitute equity do not preclude the widespread exercise of civil and political rights and the sharing of burdens and benefits associated with domestic economies. Actors use different principles of equity, whether consciously or not, in competition in the political arena in domestic society. Similar confrontations and debates take place in international relations, with sometimes analogous results.

What then is equity? Here we are concerned with *social* (as opposed to legal) justice or, more specifically, *distributive* justice, which generally refers to the "fairness" or "rightness" of distributing benefits and burdens within communities.[16] Miller gives a succinct typology of this form of equity: to each according to his rights; to each according to his deserts (actions and/or personal qualities); to each according to his needs.[17] In short, equity refers to the notion that individuals ought to receive the treatment that is proper and fitting for them. Often the concern is with distributing economic benefits, but frequently the goal is to fairly distribute other things that people care about, such as political power and liberty.

Beyond the notion of distribution within some sort of social arrangement, what else might we mean by "equity"? We might start by saying that each person ought to receive his or her due, based on rights, equality, fairness, merit ("desert"), need or some other criteria, or that each person ought to receive a share of some good depending on the extent to which that distribution will contribute to some desired consequence (e.g., the utilitarian would seek to promote the greatest overall "happiness"). But equity is not concerned merely with whom has how much of what; for one person to have more than another is not intrinsically unjust, at least according to most philosophers. For us to apply conceptions of equity to any given situation there must be some relation between actors that somehow affects the distribution.[18] Thus, equity and social justice are concerned with distributing benefits and burdens that are a result of social relationships and institutions.[19]

Discussions of equity frequently involve two general categories of issues: procedural issues (dealing with how decisions are made) and consequentialist issues (the outcomes of decisions). Procedural equity requires that rights of actors be respected in decision-making, and that those affected by decisions be allowed to participate in the formulation of those decisions. Thus we might say that those affected by pollution ought to have a say in how it will be prevented or mitigated, as well as being involved in deciding who will benefit from efforts to make right past pollution. Fair procedures can lead to unjust outcomes, so we must ask how valuable it is to follow a rule even if it usually results (or did in the past) in a just outcome. In the case of procedures or process, then, the question may be "On what principles should the distribution take place?" Alternatively, consequentialist equity demands that there be a fair (however defined) actual distribution of burdens and allocation of benefits. For example, past polluters may be required to pay those who have suffered from their pollution. Thus, from the perspective of outcomes or consequences, we might ask "Who, and how many, should have how much? What would a fair distribution look like? Would it encompass striking inequalities?"[20]

Many principles of equity will be applicable to international environmental politics depending on the issues and the actors involved. They will be especially important if concerned actors have sufficient influence to air their own interpretations of international environmental equity, something that a fair sharing of decision making authority in international environmental deliberations and institutions could go a long way toward making possible for the weaker actors. Rather than thinking about these principles as merely philosophical concepts that must meet rigorous standards of logic and reasoning, we can view them as *ways of thinking about real-world problems* of global environmental change and related problems of global economic inequity and world poverty. Indeed, these concepts have been invoked, in their various forms, in international environmental negotiations.

To be sure, some might argue that the global affluent need not or ought not aid the global poor. For example, libertarians might argue that distributive outcomes resulting from the free activities of individuals are just if the initial allocation of property rights are just. Governments and international institutions are poor managers of global resources, and are inimical to the autonomy of individuals.[21] This begs the question (which many people think can be answered in the negative) of whether the prevailing distribution of global resources and property are just.[22] What is more, environmental changes can impose conditions on persons and communities that preclude them from exercising their autonomy. Perhaps most important, for purely prudential reasons we can discount many counter arguments to the ideas introduced below, because without aid the global poor will join the affluent in destroying the environment on which all depend.

The argument here is not that the affluent should impose aid on persons or countries that do not want it, although one could make a case for doing just that if environmental conditions turn sufficiently bad. Rather, what follows are some ethical arguments for why the affluent ought to take the hard steps to aid the poor (e.g., new funding, concessional technology transfers, and the like) that might entice the less affluent persons and communities to act in the interests of all—where without these incentives their calculations of short- and medium-term interest would lead them to continue to develop in environmentally unsustainable ways. While the primary objective of this chapter is to illuminate the meaning of "international environmental equity," presenting the following principles of equity also builds the case for having more of it, thereby helping us understand why governments the world over—including the U.S. government—have begun to take it more seriously, even if they have not gone very far in acting upon it.

Principles of International Equity

Among the many specific principles of international equity, six are briefly considered below.[23] Each gives a different, but often mutually supportive, answer to the question "Why, given global environmental changes, ought the world's rich give aid to the world's poor?" Each also helps illuminate the meanings of international environmental equity.

Maximizing Human Happiness

With regard to international equity, the utilitarian might argue that any given distribution of resources should be justified based on the total amount of happiness (or "utility") it produces, measured by summing up the happiness experienced by individuals. Well-being is the only thing that is intrinsically good; it should be maximized. For utilitarians, principles of equity are rules for distributing resources that have been shown to bring beneficial results (i.e., a great deal of aggregate happiness). Thus the utilitarian might look at the consequences for promoting overall happiness of distributing goods based on rights, desert, need and so forth.[24] Ellis suggests that classical utilitarian philosophers, such as Jeremy Bentham, were fundamentally cosmopolitan, believing that people are simultaneously citizens of their own nations and the world, with duties to the good of humankind in general.[25] Bentham wrote that it is criminal for a country "to refuse to render positive services to a foreign nation, when rendering of them would produce more good to the last mentioned nation, then it would produce evil to itself."[26]

Singer describes a cosmopolitan interpretation of utilitarian equity in his seminal essay, "Famine, Affluence, and Morality."[27] His argument, in short, is that if it is in our power to prevent something very bad from happening, without thereby sacrificing anything that is morally significant, we have a moral duty to do it. And, because we are able to do so, we ought to act to prevent, for example, the starvation of thousands of people outside our society, even if this means sacrificing (for example) the upholding of property norms in our own society. Most people in the affluent countries ought to go to great lengths to help those abroad, certainly sacrificing much more at home than they are at present; "we ought, morally, to be working full time to relieve great suffering of the sort that occurs as a result of famine or other disasters."[28]

The problem with utilitarianism, we are told, is that its goal to achieve a distribution leading to the greatest overall utility could lead to some individuals suffering if such an outcome contributes to overall happiness.[29] This, on the face of it, seems to be unjust. But in the context of potentially severe global environmental damage resulting from human activity, it may indeed be fair to have the few suffer to save the planet on which we all depend for our wealth, health and survival.[30] It hardly seems just to let people exercise their individual rights, however defined (e.g., the right to spew carbon dioxide, a potent greenhouse gas, as we drive our Range Rovers into the far-off hills to enjoy the pleasures of recreational sports), if doing so will contribute to a slow environmental Armageddon. To use Singer's illustration regarding famine relief, "it should be clear that we would have to give away enough to ensure that the consumer society, dependent as it is on people spending on trivia rather than giving to famine relief, would slow down and disappear entirely."[31] This may seem a bit extreme, at least in the short term. But is it worth considering moving in that direction. Trading "trivia" for protection of the global environment, should it become necessary—as it will if we do not take effective action soon—seems a reasonable deal. So we ought not reject utilitarianism outright; utilitarian principles of international equity ought to at least be considered in the context of global environmental change.

From the utilitarian perspective we might say that individuals, with the assistance and encouragement of governments, ought to take action to promote what we might call the utility of environmentally sustainable development. This could mean that the affluent countries ought to be aiding the poor countries to achieve sustainable development, because to do so would simultaneously reduce human suffering (and thereby increase overall "utility") and reduce—and potentially reverse—environmental destruction, which could otherwise minimize happiness in the future. Utilitarianism compels us to look at overall consequences of our individual actions and suggests that it may be appropriate to treat some individuals inequitably (by other interpretations) if the environment vital to all life will be protected as a consequence.[32] The advantage of this perspective is that national boundaries are (conceptually) little impediment; the goal is utility maximization, which is going to be greater on a global scale than on a national one. This requires one to find an accurate and agreeable measure of utility that can be used to allocate benefits and burdens—a difficult task indeed, at least regarding maximizing the utility of costs and benefits associated with addressing complex environmental issues like stratospheric ozone depletion and climate change. But this difficulty should not preclude us from thinking in these terms and trying to maximize utility defined in terms of environmentally sustainable development.

Promoting Human Rights

There is a well established tradition of trying to apply conceptions of rights to international relations, but historically rights have been held almost exclusively by states. However, contemporary notions increasingly encompass the rights of individuals.[33] From the perspective of human rights, one might say that individuals have inherent rights to minimum nutrition, freedom from torture, freedom of expression, and so forth, simply because they are human beings. At the very least, we might say that individual persons ought to have their security and subsistence rights (however defined) respected, for without those rights all others cannot be fulfilled.[34] Rights can be difficult to apply in practice to international relations because their establishment often begs the questions of who or what (e.g., individuals, peoples, governments) is entitled to what rights and who or what is responsible for fulfilling those rights. Increasingly the international community is recognizing the rights of peoples and individuals, and it is now willing to occasionally violate state sovereignty to ensure those rights, as demonstrated most profoundly in the NATO intervention in Kosovo. So far, however, such action has not been taken to promote the right to development or a clean environment.[35]

How might we use notions of human rights to help us understand and think about global environmental change? Shue's principle of "basic rights" is useful.[36] In trying to show that something ought to be a right, we need to show that it is vital—either literally necessary to, or highly valuable to, living as a human—and that it is vulnerable (subject to widespread threats that individuals on their own often could not defend against).[37] At the very least, humans have a right to survival, for without life all other rights cannot be exercised. Shue says that there are duties attached to human rights: basic human rights require actors to avoid depriving, to protect from deprivation, and to aid the deprived. Depending on the circumstances, these duties may be ascribed to different actors (e.g., individuals, non-governmental organizations, state governments), although states, by virtue of their capacities, are the most important duty-bearers, and the nature of the duty may vary (e.g., to avoid depriving individuals of the right to subsist, to protect refugees from assault, or to bring aid in times of natural disaster).[38]

Insofar as human-caused pollution and resource exploitation deny people or entire communities and even countries the capacity to achieve survival, individuals who consume and pollute more than necessary have some obligation to stop those activities now underway. Also, governments might actively try to stop emissions of pollutants from within their jurisdictions that harm people in other countries. Shue's argument might also support an obligation to actively try to help people face the challenges of environmental change, to help them implement their own sustainable development measures and to help them deal with the consequences of, say, pollution of regional seas or the atmosphere. To be sure, it would be difficult to identify obligations to attach to the environment-subsistence connection. (Who caused how much pollution taking away how much of which person's or which group's ability to survive?) But just because it is difficult to identify the details of how to operationalize rights in this regard does not mean that it is not worth considering in international deliberations on environmental issues. In specific cases it may be possible to identify who is harming whom, and what activities they ought to stop and what they should do to help those they have harmed. At least the serious disputation of these issues will put polluters on notice that they are probably violating some person's or some country's rights, and that they may be sanctioned in the future as a consequence.[39]

Helping the Poor to Be Free Agents

Kantians attempt to create principles of obligation that can be used as guides for action, rather than precise rules that fit all possible situations. According to a Kantian conception of ethics, all humans have obligations toward one another by virtue of their common humanity. The Categorical Imperative requires that people ought not be treated as means to one's own ends, but as ends in themselves.[40] According to O'Neill, "We use others as *mere means* if what we do reflects some maxim *to which they could not in principle consent.*"[41] Thus it is unethical to exploit other persons or to deny their human rights. Distinctions between right and wrong are known by instinct; moral principles are *a priori*, and because by definition those principles have the consent of those affected by them, they should be followed.

Kantian principles of human equality and right and wrong action might be used to assess obligations of people in different countries toward one another. "Categorical imperatives" could be established for international relations. Duties of equity, by the Kantian conception, exist whenever there is involvement between actors, something that is common and widespread in the modern world.[42] O'Neill believes that the common humanity shared by all people requires that we meet basic minimal obligations toward one another, with these obligations deriving from basic Kantian assumptions, such as truth telling and non-coercion in social relations. But this begs the question of capacity. If we are all to treat one another as ends, we must each have the capacity to do so; "our obligations as moral agents require us to help others to be moral agents."[43] The people of the affluent countries, therefore, ought to start by assisting the people of the poor countries to become Kant's moral agents.[44] Thus in a discussion of ending world hunger, O'Neill declares that:

> Since hunger, great poverty, and powerlessness all undercut the possibility of autonomous action, and the requirement of treating others as ends in themselves demands that Kantians standardly act to support the possibility of autonomous action where it is most vulnerable, Kantians are required to do what they can to avert, reduce, and remedy hunger. They cannot of course do everything to avert hunger: but they may not do nothing.[45]

Extending this argument to global environmental issues, citizens of all countries might be obligated to refrain from unsustainable use of natural resources or from pollution of environmental commons shared by people living in other countries—or at least be obligated to make a good effort toward that end. (This begs the question of poverty alleviation, which is necessary for fulfilling such obligations in poor countries.) Insofar as the actions of individuals, corporations and governments contribute to the effects of global environmental change, if those effects are imposed on people or communities without their consent and/or prevent or limit those others' abilities to act as moral agents, we can say that the actions are a violation of Kantian interpretations of international equity. Affluent countries need not return to the Stone Age when fulfilling these obligations. What this conception seems to demand, rather, is that these considerations be taken seriously in decision making. Kantian ethics aim to "offer a pattern of reasoning by which we can identify whether *proposed action or institutional arrangements* would be just or unjust, beneficent or lacking in beneficence. . . The conclusions reached about particular proposals for action or about institutional arrangements will not hold for all time, but be relevant for the contexts for which action is proposed."[46] Seriously considering Kantian conceptions of equity in international environmental relations, and trying to find ways to make domestic action and foreign policy fit these conceptions, might go a long way toward promoting international equity. Inequity, in the environmental area and in all others, will never be eliminated, but that should not preclude efforts to minimize it.

Righting Past Wrongs

According to this conception of international equity, those not responsible for causing a problem (e.g., pollution) should not have to pay to fix it, and those responsible for causing harm are at least responsible for righting it.[47] Shue argues that "the obligation to restore those whom one has harmed is acknowledged even by those who reject any general obligation to help strangers. . . . one virtually always ought to 'make whole,' insofar as possible, anyone whom one has harmed. And this is because one ought even more fundamentally to do no harm in the first place."[48] Put another way, "If it is the case that the poverty [or, we could add, adverse environmental change] of poor countries/peoples is the result of actions by rich countries/peoples, then there would seem to be quite a strong *prima facie* case for saying that the latter have a clear responsibility to act in such a way as to make reparations. . . ."[49]

There seems to be little doubt that the developed industrialized countries are inordinately responsible for global environmental pollution. One-quarter of the world's population lives in the developed countries, yet they use three-fourths of the world's goods and services. The United States, with one-twentieth of the world's population, produces about one-quarter of the world's greenhouse gases, and the per capita emissions of greenhouse gases in the United States are nine times the global average.[50] Energy use per capita in the industrialized countries is over thirty times that in all developing countries—and one hundred times that in the least developed countries![51] If we accept that the affluent countries are indeed responsible for a disproportionate share of (or even most) global pollution, then the debate can move to questions of what to do about it. The so-called "polluter pays principle" is an accepted standard in the affluent countries.[52] Interpretations of international equity based on causality and responsibility suggest that this principle ought to be applied among national communities, not just within them.

Determining responsibility for environmental pollution is often very difficult, especially when we are addressing issues as complex as climate change. The affluent countries as a group—and the United States in particular—deserve the bulk of the blame for the problem, but assessing which affluent countries have caused how much harm to which poor—or other affluent—countries is a daunting task indeed, especially in light of rapidly increasing emissions of greenhouse gases from the poor countries.[53] Nevertheless, what this interpretation of international environmental equity calls for is, at least, that "Poor nations ought not be asked to sacrifice in any way the pace or extent of their own economic development in order to help to prevent the climate [or other environmental] changes set in motion by the process of industrialization that has enriched others."[54] As always, this is something that is and should be taken into serious consideration and acted upon; it is not necessarily a description of a future world. Serious discussions of the consequences of the North's pollution, of who precisely is responsible for suffering in which countries, and similar questions, would be signs that considerations of international environmental equity are being taken seriously.

Aiding the World's Least Advantaged

Rawls develops a contractual theory of social justice, largely in response to the deficiencies he finds in utilitarianism.[55] According to Rawls, individuals would choose core principles of justice: the equal liberty principle and the difference and fair opportunity principles. The latter seem most useful for thinking about international equity and environmental change.[56] According to the difference and fair opportunity principles, social and economic inequalities are to be arranged so that they are both:

(a) to the greatest benefit of the least advantaged, consistent with the just savings principle [the difference principle], and

(b) attached to offices and positions open to all under conditions of fair equality of opportunity [the fair opportunity principle].[57]

Inequalities in distribution are acceptable insofar as they benefit the least advantaged in society, and, even more important, they should be arrived at based on equality of opportunity. Rawls justifies these principles by saying that they accord with our intuitive moral judgments (they are "common sense") and rational individuals would agree on principles of a just society if they were to make their choice in the "original position" behind a "veil of ignorance," where they would not know their position in society, their skin color, their religion, and so forth.

Beitz applies Rawls's theory to international relations because international society is sufficiently cooperative and interdependent to be classified as a "community."[58] According to Beitz, global interdependence has created a situation in which national societies are neither self-contained nor self-sufficient, and,

> if evidence of global economic and political interdependence shows the existence of a global scheme of social cooperation, we should not view national boundaries as having fundamental moral significance. Since boundaries are not coextensive with the scope of social cooperation, they do not mark the limits of social obligations. Thus the parties to the original position cannot be assumed to know that they are members of a particular national society, choosing principles of justice primarily for that

society. The veil of ignorance must extend to all matters of national citizenship, and the principles chosen will therefore apply globally.[59]

Rawlsian principles seem applicable to issues of global environmental change; such issues are gradually driving countries together in cooperative efforts to prevent mutually destructive consequences, and it is the least advantaged in international society—the least developed countries of Africa, Asia and Oceania—that are most vulnerable to environmental pollution (and its effects) originating in more affluent places. In those instances where the environmental plight of small, weak countries are being taken seriously by the international community—such as the small island states that are particularly vulnerable to the effects of climate change (namely, sea-level rise)—we might say that Rawls's concern for the least advantaged has been operationalized in international environmental relations, or at least there is a move in that direction. Furthermore, the equal opportunity principle that is part of Rawls's theory of justice, if applied to international relations, suggests strongly that diplomats should give serious consideration to equal opportunity in cooperative arrangements—international institutions—directed toward protecting the global environment. That means giving the poor countries more of a say in decisions regarding the sharing among countries of benefits and burdens associated with environmental change.[60]

Being Impartial

According to a conception of justice developed by Barry, actors have interests which sometimes come into conflict, but they use the "arsenal of persuasion" to arrive at agreements on terms that no one of them could reasonably reject.[61] The motive for behaving justly is, according to Barry,

> the desire to act in accordance with principles that could not reasonably be rejected by people seeking agreement with others under conditions free from morally irrelevant bargaining advantages and disadvantages. . . . The significance of speaking of "justice as impartiality" is that this approach, however it is worked out in detail, entails that people should not look at things from their own point of view alone but seek to find a basis of agreement that is acceptable from all points of view.[62]

The most important question is "What is reasonable?" The goal of mutual advantage as the underlying goal of cooperation is replaced by a desire to reach agreement. According to Barry, it will be impossible to persuade actors to negotiate based on self-interest and reciprocity if those actors do not wish to reach a reasonable agreement.[63] Like Beitz (but using different reasoning), Barry envisions substantial redistributions between countries to ensure that each has a fair share of the world's resources. Countries would agree to such transfers because they are reasonable, not because they expect a reciprocal response. In Brown's words: "As between the United States and Bangladesh, there can be no reciprocity—for the foreseeable future the relationship will be one-way. The United States should aid Bangladesh not because it is in the United States' interest to do so but because justice as impartiality suggests that the case for such aid cannot be reasonably denied."[64]

Countries should ask themselves what is reasonable to expect of one another. For example: Is it reasonable to continue to emit pollutants that contribute to climate change and that will have especially adverse effects in the poorest countries, and is it reasonable to deny poor countries the help they require to cope with climate change and to join in efforts to prevent it, especially when the North is disproportionately at fault? If statespersons refuse to be subject to such judgments, then considerations of international equity are not possible. Paterson suggests that Barry's framework offers the most convincing grounding for equity in the context of climate change, and Barry's argument would justify greater egalitarianism between countries, including transfers between North and South to address climate change.[65] Increasingly, statespersons are entering international environmental negotiations with the assumption that agreement is achievable if all participants treat one another fairly and impartially, insofar as doing so does not pose an undue strain on their own important short-term national interests.[66]

Conclusion

We return to the original question: What is international environmental equity? While the precise answer to this question will always be a result of political bargaining among actors involved in international environmental deliberations, we can at least say what equity is about in this context (as it is in others). It is about giving serious consideration to the condition of all people everywhere, and trying to promote the well-being of all. It is about those who are doing harm to others stopping that harm, or if the harm is not ultimately harmful to the planet at the very least paying some form of restitution to those harmed (but only if that is what they choose). It is about meeting the needs of the world's poorest people, and ensuring that in the future their basic needs—their basic rights—do not suffer from present and future policies chosen by others. It is about letting all countries, including those that are poor and weak, have a say in their and the world's future. It is about, perhaps most fundamentally (and perhaps tautologically, at least in this discussion), treating all countries fairly and impartially. International environmental equity is about all countries—especially the rich developed countries in their relations with the poor developing countries—given serious consideration to fairness and justice in global environmental politics. That this is not a particularly concrete definition simply demonstrates the challenges faced by diplomats.

I will argue in the following chapters that there is great practical significance in integrating equity into global environmental policy making. But here I have tried to suggest that there are strong ethical arguments, from a variety of philosophical perspectives, for promoting international environmental equity. While governments will continue to act primarily to promote their own national interests, ethical arguments can redirect their policies when their vital short-term interests are not threatened. Considerations of international equity can bolster these policies, and thinking about them can help scholars and laypersons understand the problems faced by negotiators at international environmental negotiations.

Environmental issues, more than any others, are compelling governments and diplomats the world over—including those of the affluent countries—to seriously consider international equity. Questions of equity and justice in international relations were common in the 1970s—and subsequently declared moribund by most of the affluent countries in the 1980s. Yet, as the global environment becomes more polluted in the future, considerations of international equity will become increasingly germane to international relations. Environmental change and closely related requirements for sustainable development and poverty eradication in the developing world have already pushed equity back onto the global political agenda. This movement toward more serious consideration of equity in the context of global environmental politics, culminating in the Earth Summit and related international environmental deliberations, is the subject of the next two chapters.

Notes

1   I first discussed these ideas in greater detail in Paul G. Harris, "Affluence, Poverty and Ecology: Obligation, International Relations and Sustainable Development," *Ethics and the Environment* 2, 2 (Fall 1997), pp. 121-38.

2   For an elaboration of this definition, see Paul G. Harris, "Defining International Distributive Justice: Environmental Considerations," *International Relations* 15, 2 (August 2000), pp. 51-66.

3   David Miller, *Social Justice* (Oxford: Clarendon Press, 1976), p. 18.

4   "Foul" weather in England may mean heavy rain; in sub-Saharan Africa it may mean lack of rain. In both cases it may be caused by what people do half-way around the world. For the most respected analyses of climate change and the potential human causes of it, see J.J. Houghton et al., eds., *Climate Change 1995: The Science of Climate Change* (New York: Cambridge University Press, 1996).

5   Assuming the scientists are correct in their dire—and increasingly well informed—predictions. I will operate under the assumption that the scientists' forecasts are essentially accurate. See Houghton et al.

6   For an examination of equity in the context of climate change, see Ferenc L. Toth, ed., *Fair Weather?: Equity Concerns in Climate Change* (London: Earthscan Publications, 1999).

7   Kjell Tornblom, "The Social Psychology of Distributive Justice," in Klaus R. Sherer, ed., *Justice: Interdisciplinary Perspectives* (Cambridge, England: Cambridge University Press, 1992), p. 177.

8   Bernard Cullen, "Philosophical Theories of Justice," in Sherer, p. 60.

9   Hedley Bull, *The Anarchical Society* (New York: Columbia University Press, 1977), p. 78.

10 Stanley Hoffman, *Duties Beyond Borders* (Syracuse: Syracuse University Press, 1981), p. 143.

11 Even within the most economically developed societies, debates about how to distribute economic and other resources are ongoing and frequently heated. We should not expect *international* affairs to be any different. I reiterate this because so many people seem to expect miracles from international relations that are difficult or impossible to achieve even in small, relatively harmonious countries.

12 Near-universally accepted examples of specific injustices might include genocide, torture, intentional starvation, widespread use of atomic weapons, involuntary exploitative colonialism, intentional destruction of large portions of the natural environment, and so forth.

13 Miller, p.28. Accepting this diversity does not preclude having preferences among them.

14 Janna Thompson, *Justice and World Order: A Philosophical Inquiry* (New York: Routledge, 1992).

15 See Robert D. Bullard, ed., *Unequal Protection: Environmental Justice and Communities of Color* (San Francisco: Sierra Club Books, 1994); Richard Hofrichter, *Toxic Struggles: The Theory and Practice of Environmental Justice* (Philadelphia: New Society Publishers, 1993); and Evan J. Ringquist, "Environmental Justice: Normative Concerns and Empirical Evidence," in *Environmental Policy in the 1990s*, eds. Norman J. Vig and Michael E. Kraft (Washington: CQ Press, 1994), pp. 232-256 (and works cited therein).

16 This, of course, begs the question of precisely what is fair and right, to which I turn very briefly later, without coming to any resolution. In short, what attributes are attached to "justice" are very often, for better or worse, the result of political bargaining among interested actors.

17 Miller, pp. 25-121. Clearly, there will be conflicts between these interpretations, as Miller acknowledges.

18   John Arthur and William H. Shaw, *Justice and Economic Distribution* (Englewood Cliffs: Prentice Hall, 1978), pp. 2-8.

19   Miller, p. 22.

20   Brenda Almond, "Rights and Justice in the Environment Debate," in David E. Cooper and Joy A. Palmer, eds., *Just Environments: Intergenerational, International and Interspecies Issues* (New York: Routledge, 1995), p. 12.

21   Brendan O'Leary, "Libertarianism," in Kenneth McLeish, ed., *Key Ideas in Human Thought* (London: Clays, 1993), p. 425; cf. Robert Nozick, *Anarchy, State and Utopia* (New York: Basic Books, 1974).

22   A libertarian approach would maintain the status quo, which is increasingly bad for the environment and for humans, especially the poorest humans.

23   The general typology of justice used here is derived from Matthew Paterson, "International Justice and Global Warming" (paper for Conference on Ethics and Global Change, Reading University, 29 October 1994), pp. 10-16, and appears in Paul G. Harris, "Considerations of Equity and International Environmental Institutions," *Environmental Politics* 5, 2 (Summer 1996), pp. 274-301. Paterson derives the bulk of his framework from Chris Brown, *International Relations Theory: New Normative Approaches* (New York: Columbia University Press, 1992).

24   Miller, p. 32.

25   Anthony Ellis, "Utilitarianism and International Ethics," in Terry Nardin and David R. Mapel, eds., *Traditions of International Ethics* (Cambridge: Cambridge University Press, 1992), p. 164.

26   Jeremy Bentham, "Principles of International Law" in John Bowring, eds., *The Works of Jeremy Bentham*, vol. 2 (New York: Russell and Russell, 1962), p. 538.

27   Peter Singer, "Famine, Affluence, and Morality" in William Aiken and Hugh LaFollette, eds., *World Hunger and Morality* (Upper Saddle River: Prentice Hall, 1996).

28   We might insert ozone depletion and especially climate change for Singer's "other disasters". Singer, p. 33.

29   Almond, p. 12.

30   In this scenario, the question of who the "few" are is central. Some might (and have) said that population growth in the poor countries is the chief cause of environmental destruction; others have blamed it on consumption patterns in the affluent areas of the global North. Perhaps reasonable but traditionally "unjust" constraints on individuals in developed economies—restrictions on energy use, regulations mandating recycling, and the like—would be the way to go, or at least a start, as opposed to trying to hold down the already destitute in the developing world. As Shue puts it, "whatever justice may positively require, it does not permit that poor nations be told to sell *their* blankets in order that rich nations may keep *their* jewelry." Henry Shue, "The Unavoidability of Justice," in Andrew Hurrell and Benedict Kingsbury, eds., *The International Politics of the Environment* (New York: Oxford University Press, 1992), p. 397.

31   Singer, p. 36.

32   Individual actions are emphasized, something that we in the wealthy countries frequently avoid thinking about. The "Think Globally, Act Locally" bumper stickers are one effort to get individuals in the affluent countries to restrain their individual "utility maximization" (by polluting less, recycling more, and so forth) in order to benefit the global environment and all species that rely on it.

33 See the works cited in R.J. Vincent, *Human Rights and International Relations* (Cambridge, England: Cambridge University Press, 1986) and Jack Donnelly, *Universal Human Rights in Theory and Practice* (Ithaca: Cornell University Press, 1989; 1993).

34 Henry Shue, *Basic Rights*, 2nd. ed. (Princeton: Princeton University Press, 1996).

35 The right to development was codified in the 1986 United Nations Declaration on the Right to Development. Dower suggests that there should also be a right to *sustainable* development. See Nigel Dower, "Sustainability and the Right to Development," in Robin Attfield and Barry Wilkins, eds., *International Justice and the Third World* (New York: Routledge, 1992).

36 Shue, *Basic Rights.*

37 Henry Shue, "Solidarity Among Strangers and the Right to Food," in William Aiken and Hugh LaFollette, eds., *World Hunger and Morality* (Upper Saddle River: Prentice Hall, 1996), p. 114.

38 R.J. Vincent, "The Idea of Rights in International Ethics," in Terry Nardin and David R. Mapel, eds., *Traditions of International Ethics* (Cambridge, England: Cambridge University Press, 1992), p. 255, 260.

39 Such a prospect may become more likely as procedures and technologies to make such assessments become available and cost-effective.

40 Immanuel Kant, *The Moral Law* (London: Hutchinson, 1948). This "Formula of the End in Itself" is but one of Kant's formulations of the Categorical Imperative. See Onora O'Neill, "Kantian Ethics," in Peter Singer, ed., *A Companion to Ethics* (Oxford: Blackwell, 1993).

41 Onora O'Neill, "Ending World Hunger" in William Aiken and Hugh LaFollette, eds., *World Hunger and Morality* (Upper Saddle River: Prentice Hall, 1996), p. 97.

42 Ibid., p. 103.

43 Brown, p. 170.

44 Onora O'Neill, *Faces of Hunger: An Essay of Poverty, Development and Justice* (London: Allen and Unwin, 1986).

45 O'Neill, "Ending World Hunger," p. 99.

46 Ibid., p. 103.

47 I say "at least" because one might suggest that those responsible for causing harm ought to first correct it and second pay some sort of fine as punishment—an additional incentive not to cause harm in the future.

48 Henry Shue, "Equity in an International Agreement on Climate Change" in Richard Samson Odingo et al., eds., *Equity and Social Considerations Related to Climate Change* (Nairobi: ICIPE Science Press, 1995), p. 386; see also Henry Shue, "Subsistence Emissions and Luxury Emissions" *Law and Policy* 15, 1 (January, 1993), pp. 39-59.

49 Brown, p. 159.

50 United Nations Development Program (UNDP), *Human Development Report 1995* (New York: Oxford University Press, 1995). These trends did not change much in the subsequent six years.

51 Ibid., p. 191.

52 The OECD countries agreed to base their environmental policies on the polluter pays principle in 1972.

53 One especially difficult question surrounds assessing responsibility for historical emissions of greenhouse gases and other pollutants (as opposed to present and future emissions, which will be difficult enough). How responsible are the developed

countries for their historical emissions, which they did not know would harm other countries? This raises the paradoxical notion that the developing countries may have a higher standard of duty than did the North for much of its history. The developing countries know that fossil-fuel intensive industrialization will harm others. However, the South may lack the capacity to act on such notions without assistance from the North.

54 Shue, "The Unavoidability of Justice" p. 395.

55 Here I focus narrowly on portions of Rawls's argument, especially his "difference principle." John Rawls, *A Theory of Justice* (Cambridge, MA: Harvard University Press, 1971).

56 The equal liberty principle states that "Each person is to have an equal right to the most extensive total system of equal basic liberties compatible with a similar system of liberty for all." Ibid., p. 302.

57 Ibid.

58 Charles Beitz, *Political Theory and International Relations* (Princeton: Princeton University Press, 1979). Rawls says that if countries were to come together in an "original position" they would choose principles of justice familiar to international lawyers: equal rights of states, self determination and non-intervention, right to self-defense, justice of and in war (*jus ad bellum, jus in bello*) and the like. Rawls, pp. 377-82 and passim.

59 Beitz, *Political Theory and International Relations*, p. 151. Beitz subsequently changed his reasoning. While he thinks that Rawls's difference principle still ought to apply to international relations, he bases this conclusion on a Kantian perspective, namely that those eligible for participation in the original position need only possess the two essential powers of moral personality: a capacity for an effective sense of justice and a capacity to form, revise and pursue a conception of the good. Charles Beitz, "Cosmopolitan Ideals and National Sentiment," *Journal of Philosophy*, 80 (1983), p. 595. For a more recent treatment of this issue by Beitz, see Charles Beitz, "Sovereignty and Morality in International Affairs," in David Held, eds., *Political Theory Today* (Stanford: Stanford University Press, 1991).

60 This is what has happened in the Global Environment Facility and the Montreal Protocol's Multilateral Fund.

61 Brian Barry's ideas, developed in several works, are summarized in Brown. See Brian Barry *Theories of Justice* (Berkeley: University of California Press, 1989) and *Justice as Impartiality* (Oxford: Clarendon Press, 1995).

62 Barry, *Theories of Justice*, p. 8. This is analogous to Kantian conceptions of justice, as Barry acknowledges.

63 See also Brown, pp. 180-81.

64 Ibid.

65 Paterson, p. 19.

66 Norwegian United Nations ambassador Sven Aass said as much during a speech on sustainable development at Brandeis University, 23 March 1996. Aass is supported by the author's discussions with several delegates and their deputies involved in the United Nations Intergovernmental Negotiating Committee (INC) on Climate Change, New York, 9-10 February 1995, including: Rafe Pomerance, U.S. Deputy Assistant Secretary of State for Environment and Development, interview by author, New York, 10 February 1995.

3   The Earth Summit and International Equity

Provisions for international equity were pervasive throughout the agreements and conventions signed at the Earth Summit in 1992, as well as the statements made there. More than any previous international conference or agreement, UNCED and the Earth Summit contributed to the codification of rights and obligations of international equity, at least insofar as they apply to the environment and sustainable development. The Earth Summit marked a turning point at which poverty, economic development, and environmental change became inextricably linked in international discourse. The results so far have been mixed and rather disappointing, but the normative foundation has been laid.

The Earth Summit showed that considerations of equity could not be ignored by the developed countries to the degree that they were in the 1970s during calls for a New International Economic Order. Moreover, the provisions for equity in UNCED agreements and conventions go beyond those of the 1972 UN Conference on the Human Environment (UNCHE), the 1982 UN Convention on the Law of the Sea (UNCLOS), and the 1987 Montreal Protocol on Substances that Deplete the Ozone Layer. (The equity provisions of UNCHE, UNCLOS, the Montreal Protocol, and other international environmental agreements, are described in the next chapter.) Equity considerations continued to be prominent in the UNCED follow-on negotiations, notably those regarding the Climate Change Convention.

This chapter describes many of the provisions for international equity that permeate the UNCED agreements and conventions signed at the Earth Summit, as well as other UNCED products and follow-on arrangements still being deliberated. The chapter concludes with a brief examination of ways in which those provisions for international environmental equity resemble arguments made by political philosophers.

The Earth Summit and the Rio Declaration

One hundred seventy-eight countries and 118 heads of state participated in the Earth Summit, making it the largest international conference to that date. Many official statements and agreements that UNCED produced, if faithfully implemented by the signatories, would promote international equity. Products of the UNCED process that were agreed at the Earth Summit included the Rio Declaration on Environment and Development; Agenda 21, the lengthy policy statement of UNCED; conventions on climate change and biological diversity; a nonbinding statement on forest principles; establishment of a commission on sustainable development (CSD); expansion of the Global Environment Facility (GEF); as well as increased emphasis on environmental awareness in the World Bank, the International Monetary Fund (IMF), the UN Development Program (UNDP), and other UN agencies. To varying degrees, all of these UNCED products incorporated significant equity provisions (compared to other international agreements), such as calls for concessionary or preferential technology transfers, soft loans, and new and additional funding for developing countries.

The Rio Declaration on Environment and Development contains several provisions for international equity.[1] It acknowledges that individuals are the center of concerns for sustainable development[2] and that "Peace, development and environmental protection are interdependent and indivisible."[3] The Declaration states in Principle 3 that "The right to development must be fulfilled so as to equitably meet developmental and environmental needs of present and future generations" and Principle 5 declares that "All States and all people shall cooperate in the essential task of eradicating poverty as an indispensable requirement for sustainable development, in order to decrease the disparities in standards of living and better meet the needs of the majority of the people of the world." Furthermore, the countries at Rio declared that the "special situation and needs of developing countries, particularly the least developed and those most environmentally vulnerable, shall be given special priority."[4] Principle 7 declares that states have "common but differentiated responsibilities," meaning that developed countries have a greater responsibility to take steps to protect the global environment—and to help less affluent countries do likewise. Thus the Declaration states that "The developed countries acknowledge the responsibility that they bear in the international pursuit of sustainable development in view of the pressures their societies place on the global environment and of the technologies and financial resources they command."[5]

Agenda 21

Agenda 21, the comprehensive UNCED policy statement adopted at Rio, promotes equity in most of its provisions, with several chapters concentrating on such issues (particularly Chapters 2, 3, 33 and 34). For example:

> The developmental and environmental objectives of Agenda 21 will require a substantial flow of new and additional financial resources to developing countries, in order to cover the incremental costs for the actions they have to undertake to deal with global environmental problems and to accelerate sustainable development.[6]

> The struggle against poverty is the shared responsibility of all countries.[7]

> The implementation of the huge sustainable development programmes of Agenda 21 will require the provision to developing countries of substantial new and additional financial resources. Grant or concessional financing should be provided according to sound and equitable criteria and indicators. The progressive implementation of Agenda 21 should be matched by provision of such financial resources. The initial phase will be accelerated by substantial early commitments of concessional funding.[8]

> For developing countries, particularly the least developed countries, ODA [official development assistance] is a main source of external funding, and substantial new and additional funding for sustainable development and implementation of Agenda 21 will be required. Developed countries reaffirm their commitments to reach the accepted United Nations target of 0.7 per cent of GNP [gross national product] for ODA and, to the extent that they have not yet achieved that target, agree to augment their aid programmes in order to reach that target as soon as possible and to ensure prompt and effective implementation of Agenda 21.[9]

> [The nascent Global Environment Facility, which was expected to fund many Agenda 21 programs] should be restructured so as to *inter alia*: Encourage universal participation; . . . Ensure a governance that is transparent and democratic in nature, including in terms of decision-making and operations, but guaranteeing a balanced and equitable representation of the interests of developing countries. . .; Ensure new and additional financial resources on grant and concessional terms, in particular to developing countries; Ensure predictability in flow of funds by contributions from developed countries. . .; Ensure access to and disbursement of the funds under mutually agreed criteria without introducing new forms of conditionality.[10]

> [Parties to Agenda 21 should] promote, facilitate, and finance, as appropriate, the access to and the transfer of environmentally sound technologies and corresponding know-how, in particular to developing

countries, on favorable terms, including on concessional and preferential terms. . . .[11]

Funding Agenda 21 programs was estimated by the UNCED secretariat to require approximately $600 billion annually, with $125 billion of that coming from the developed countries.[12] The latter amount was about $70 billion more than the total ODA at the time of the Earth Summit.[13]

Forest Principles

UNCED also produced a nonbinding statement of forest principles that calls on countries to minimize damage to their forests and undertake programs to determine how economic development affects their forests.[14] The North wanted to protect the world's forests because trees can act as sinks for greenhouse gases (at least temporarily) and contain much of the world's genetic resources that North-based corporations develop into pharmaceuticals and other biotechnologies. Leaders of the Group of Seven (G-7) industrialized countries agreed at their 1990 meeting in the United States to push for the negotiation of a treaty on the world's forest "as expeditiously as possible."[15] President Bush was perhaps the strongest backer of such an effort.[16] However, several developing countries—notably Malaysia, Indonesia, and Brazil—feared that such a treaty would limit their freedom to exploit their forests, thereby constraining their ability to earn scarce financial resources for economic development. Malaysia was emphatic throughout the negotiations that it would never renounce its right to cut its forests if the developed countries would not compensate it fully for doing so.[17] Developing countries also wanted compensation for genetic resources used by multinational corporations (usually from the developed countries) that come from Southern forests. As a result of these and other differences, diplomatic efforts to negotiate a forest treaty were abandoned, replaced instead with "A Non-Legally Binding Authoritative Statement of Principles for a Global Consensus on the Management, Conservation and Sustainable Development of All Types of Forests" (as it was called officially).[18]

One of the U.S. delegates involved in the early UNCED negotiations commented on the U.S. failure to negotiate a forest convention, which the United States was aggressive in promoting during 1990, this way:

The U.S. initiative was derailed by a combination of developing country perceptions that the U.S. was trying to shift responsibility for its own fossil fuel related carbon dioxide emissions to sink enhancement in developing countries, and that the U.S. was not going to marshal significant resources to assist in protecting forests. Further, it was considered inequitable to call for developing country action on forests while the U.S. was (and is) still cutting its own ancient growth forests, and the industrialized countries had already largely cut down their original forests.[19]

The statement of forest principles declares that each country has the sovereign and inalienable right to utilize, manage and develop its forests in a manner that fits with its needs for socioeconomic development. The statement calls for the "eradication of poverty"[20] as well as "new and additional resources"[21] and "access to and transfer of environmentally sound technologies and corresponding know-how on favorable terms, including concessional and preferential terms"[22] for the developing countries.[23]

The Climate Change Convention

In the run-up to the Earth Summit, the developing countries pointed out that the vast majority of emissions of greenhouse gases were the responsibility of the developed countries. They constantly reminded those who would listen that the industrialized countries consume 80 percent of the world's resources and contribute the most to climate change.[24] The United States alone, with only five percent of the world's population, accounts for one quarter of the world's energy consumption, mostly in the form of fossil fuels that produce carbon dioxide (a greenhouse gas) when burned. The average American uses thirty times the amount of energy that an average Indian uses; the Indian uses two percent of the electricity that the American uses.[25] Many of the adverse effects of potential climate change are expected to disproportionately affect the developing countries, but those countries do not have the resources needed to address these problems. For the developing countries to control their emissions of greenhouse gases, the reasoning went, they would have to limit their economic development because both development and emissions have been historically linked to energy use. The North sees growth of polluting industries of the South as posing the greatest danger in the future. Most countries agreed that an effective effort to limit climate change would eventually require comprehensive participation, including that of the large developing countries whose emissions are expected to increase dramatically in this century. But the developed countries were expected to act first and to help the poorer countries when their turn to act arrives.

The United Nations Framework Convention on Climate Change (FCCC or Climate Change Convention)[26] was signed at the Earth Summit by 153 countries and the European Community, including the United States. It came into effect on 21 March 1994 after being ratified by more than 50 countries.[27] The FCCC was the first international agreement incorporating the basic elements of sustainable development.[28] The Convention is a legally binding treaty in which parties agree to voluntarily limit their emissions of carbon dioxide, methane, and other greenhouse gases that contribute to unnatural warming of the earth. Because of objections made by the United States, the Convention does not contain firm targets or timetables for countries to stabilize or reduce their emissions of greenhouse gases. (Modest targets and timetables were set in the 1997 Kyoto Protocol to the FCCC.) Other developed country governments concluded that having the United States on board a weak treaty would be better than having a stronger treaty without the United States (a conclusion that some now regret). According to the convention, parties to it,

> shall adopt national policies and take corresponding measures on the mitigation of climate change, by limiting its anthropogenic emissions of greenhouse gases and protecting and enhancing its greenhouse gas sinks and reservoirs. These policies and measures will demonstrate that developed countries are taking the lead in modifying longer-term trends in anthropogenic emissions consistent with the objective of the Convention, recognizing that the return by the end of the present decade to earlier levels of anthropogenic emissions of carbon dioxide and other greenhouse gases not controlled by the Montreal Protocol would contribute to such modification. . . . [29]

The treaty was generally interpreted as a voluntary commitment by developed countries to take steps to reduce their emissions of greenhouse gases to 1990 levels by 2000. However, the other OECD countries were ready to accept this schedule as a binding commitment, but only if the United States also agreed to do so. The United States would not agree to a binding target. Indeed, President Bush made his attendance at the Rio conference contingent on any climate treaty being devoid of such concrete commitments.

Several developed country delegates said that equity considerations played and would continue to play (as indeed they have) an extremely important role in the negotiations of the FCCC and ongoing international deliberations on climate change.[30] The FCCC calls on industrialized countries to help the developing countries by providing finance and technology to meet treaty objectives. The preamble to the convention notes that most current and historical emissions of greenhouse gases have originated in the North and that "per capita emissions in developing countries are still relatively low and that the share of global emissions originating in developing countries will grow to meet their social and developmental needs," and that actions to address climate change should first consider the "legitimate priority needs of developing countries for the achievement of sustained economic growth and the eradication of poverty."

The convention states that parties should be guided by several principles: Parties should take steps to protect the planet's climate system "on the basis of equity," with developed countries taking the lead.[31] Special consideration should be given to the "specific needs and special circumstances" of developing countries.[32] All countries party to the convention should promote, and "have the right to," sustainable development.[33] Parties are called on to make new efforts to "cooperate to promote a supportive and open international economic system that would lead to sustainable economic growth and development in all Parties, particularly developing country Parties. . . ."[34]

The convention's article on commitments has numerous provisions for international equity. In several places the treaty calls on participants to provide developing countries with aid to assist them in fulfilling the convention's information and reporting requirements.[35] Article 4 calls for new and additional resources delivered from the developed countries in an "adequate" and "predictable" fashion to assist developing countries in complying with their obligations under the convention.[36] Developed countries are to take steps to "promote, facilitate and finance, as appropriate, the transfer of, or access to, environmentally sound technologies and know-how," especially to developing countries.[37] The same article declares that developing countries' effective implementation of the treaty "will depend on the effective implementation by developed country Parties of their commitments . . . related to financial resources and transfer of technology and will take fully into account that economic and social development and poverty eradication are the first and overriding priorities" of the developing countries.[38] Other paragraphs state that participants in the treaty are to fully consider the special needs of developing and least developed countries.[39]

Developing countries joined the convention only after it was agreed that their development prospects would not suffer in the process. Such an agreement included an implicit understanding that some sort of international fund would be established to compensate developing countries for the costs of participation in the convention.[40] Article 11 describes the financial mechanism envisioned to provide moneys "on a grant or concessional basis, including for the transfer of technology" to help poorer countries fulfill treaty commitments.[41] The financing body was to "function under the guidance of and be accountable to the Conference of the Parties, which shall decide on its policies, programme priorities and eligibility criteria."[42] The financial mechanisms "shall have an equitable and balanced representation" of parties to the convention "within a transparent system of governance."[43] Details of the financial mechanism were explicitly put off to a later date. The North wanted to control funds, in part to ensure their effective use. The South wanted to participate in decision-making regarding the dissemination of funds. The Global Environment Facility, jointly administered by the UN Development Program, the UN Environment Program, and the World Bank, was designated as the interim financial mechanism, with the understanding that it would be in the near future "appropriately restructured and its membership made universal. . . ."[44]

The Biodiversity Convention

The legally binding Convention on Biological Diversity was signed at the Earth Summit by 153 countries and the European Community.[45] It entered into force on 29 December 1993. The United States was the only country at the Earth Summit not signing the treaty. In negotiations, the developing countries demanded sovereign rights over their genetic resources and control of access to those resources. They wanted concessional transfer of technologies that were developed from genetic resources extracted from their jurisdictions, and they wanted concessional access to technologies that would help them preserve those resources. In addition, they insisted on fees and royalties for access to their biological resources and the creation of a fund to help them meet the provisions of any biodiversity treaty. For the North, especially the United States, the main concern was protecting the pharmaceutical and biotechnology industries' intellectual property and their access to plants, animals, microbes and the like in developing countries.[46]

Flora and fauna (often referred to in this context as "genetic resources") have traditionally been treated as resources that could be exploited by those countries and corporations with the means to do so. With the advent of advanced biotechnologies, such resources became more financially lucrative, adding stimulus to demands by poor countries for sovereignty over them. Some diplomats from the South claimed that efforts to preserve biological diversity, without attendant efforts to spread the wealth that can accrue from those resources, was an effort by the North to benefit its own industries, rather than an effort to promote the interests of all. The South viewed the convention as a component of its larger agenda to use UNCED as a vehicle for restructuring international economic relations in order to acquire resources, technologies, and market access that would enable developing countries to develop in a way that was environmentally benign and simultaneously rapid enough to meet the growing demands of their populations.[47]

The objectives of the Biodiversity Convention—compromises intended to satisfy both developed and developing countries—are the "conservation of biological diversity, the sustainable use of its components, and the fair and equitable distribution of the benefits arising out of the utilization of genetic resources, including appropriate access to genetic resources and by appropriate transfer of relevant technologies, taking into account all rights over those resources and to technologies, and by appropriate funding."[48] The convention's Preamble acknowledges that "the provision of new and additional financial resources and appropriate access to relevant technologies can be expected to make a substantial difference in the world's ability to address the loss of biological diversity," and that "special provision is required to meet the needs of developing countries, including the provision of new and additional financial resources and appropriate access to relevant technologies." The Preamble also states that parties to the convention recognize that "economic and social development and poverty eradication are the first and overriding priorities of the developing countries."

The convention obligates parties to take steps to share "in a fair and equitable way the results of research and development and the benefits arising from the commercial and other utilization of genetic resources with the Contracting Party providing such resources."[49] Similar provisions are made for technology transfer.[50] Article 16 states that technology access and transfers to developing countries "shall be provided and/or facilitated under fair and most favorable terms, including on concessional and preferential terms where mutually agreed."[51] Parties to the treaty are required to "take all practicable measures to promote and advance priority access on a fair and equitable basis by Contracting Parties, especially developing countries, to the results and benefits arising from biotechnologies based upon genetic resources provided by those Contracting Parties."[52]

Article 20 outlines provisions for financial resources. It states that the "developed country Parties shall provide new and additional resources to enable developing country Parties to meet the agreed full incremental costs to them of implementing measures which fulfill the obligations" of the convention,[53] and that the extent to which developing countries will effectively implement their commitments according to the convention will "depend on the effective implementation by developed country Parties of their commitments related to financial resources and transfer of technology and will take fully into account the fact that economic and social development and eradication of poverty are the first and overriding priorities" of the developing countries.[54] Parties to the treaty are to take "full account of the specific needs and special situation of least developed countries in their actions with regard to funding and transfer of technology."[55]

Plans to develop a financial mechanism are described in Article 21 of the Biodiversity Convention. The first paragraph states that there "shall be a mechanism for the provision of financial resources to developing country Parties for the purposes of this Convention on a grant or concessional basis. . . ." Contributions to the fund "shall be such to take into account the need for predictability, adequacy and timely flow of funds. . . ." The funding mechanism is to operate "under the authority and guidance of, and be accountable to, the Conference of the Parties" and shall operate "within a democratic and transparent system of governance."[56] The GEF was deemed the interim funding mechanism, although it was not agreeable to the South because its voting was weighted toward donor countries. Final details of the financial mechanism were put off to a post-Rio meeting of the parties where the GEF became the permanent financial mechanism. In April 1994 GEF participants agreed to changes in the structure of the GEF, including a double-majority voting system that gives developing countries greater influence in funding decisions and gives veto powers to parties from both North and South (see below).[57]

The financial provisions of the Climate Change Convention helped set precedents for the Biodiversity Convention, but the latter went further toward the demands of the developing countries.[58] As Mott points out, the Biodiversity Convention uses simpler language in describing its funding mechanism. More important, substitution of "democratic" for the Climate Change Convention's "equitable and balanced" representation "sounds decidedly more like 'one country, one vote'—a victory for developing countries, and a formulation that industrialized countries found politically hard to oppose."[59] Furthermore, the Biodiversity Convention is unique in that its Conference of the Parties (COP) has the authority to decide the amount of financial resources needed to implement the convention. In the Climate Change Convention the GEF is under the "guidance of" and "accountable to" the COP, whereas under the Biodiversity Convention it operates according to the "authority and guidance of, and [is] accountable to" the COP.[60] What is more, codification of the notion that environmental aid should be *in addition to* other forms of development aid has been an objective of the developing countries since the 1972 Stockholm environment conference. While one ought not overstate the situation, in Alan Boyle's view, "It is remarkable that after prolonged resistance [the developing countries] now have agreement on the inclusion of this obligation in the [Biodiversity] Convention."[61]

At Rio, the developing countries met with the greatest success in gaining further recognition of their sovereign rights over their biological resources. The Biodiversity Convention's requirements that access to genetic resources must be subject to the prior informed consent of the country where the collection occurs, and that access to such resources must be on mutually agreed terms, is entirely new international law.[62] This is the outcome one would expect from negotiations between states operating in an international system where sovereignty and non-intervention are the most basic and most cherished components of the system, but not necessarily the result one might expect in a system where traditional power resources determine the course of events. Like the Stockholm conference, in the Biodiversity Convention the Earth Summit recognizes that countries have sovereignty over their genetic resources.[63] But it goes further by stating that access to genetic resources is subject to "prior informed consent" of the country providing the resources, "on mutually agreed terms." Countries and corporations utilizing biological resources are also instructed to research and develop those resources, insofar as is practicable, in the country where the resources are found. The results and commercial benefits of such research and development are to be equitably shared with the country providing the genetic resources.

The Commission on Sustainable Development

In addition to the Rio Declaration, statement on forest principles, Agenda 21, Framework Convention on Climate Change, and Convention on Biological Diversity, the Earth Summit stimulated the creation of particular institutions that can contribute to greater international equity.

At its 47th session the General Assembly adopted resolution 47/191, which recommended that the UN's Economic and Social Council (ECOSOC) establish the United Nations Commission on Sustainable Development (CSD).[64] The Commission was formally established by ECOSOC on 12 February 1993. The CSD was created to operate with 53 members at the ministerial level as a functional commission within ECOSOC. Its primary objectives are to monitor progress on implementation of Agenda 21 at the national, regional, and international levels; to ensure effective follow-up of UNCED; to encourage and coordinate international cooperation on environment and development; and to assess the transfer of financial resources and technology to the developing countries, including review of developed countries' progress toward meeting the goal of providing 0.7 per cent of their GNPs for official development assistance. Beginning in 1993, the CSD held yearly sessions in New York, with the expectation that there would be a major assessment of its work and the progress on UNCED at the UN's 1997 session (see below).[65] Importantly, the CSD has equitable geographic representation among its 53 members, ensuring that developing countries have a strong bargaining position relative to the developed countries.[66] Extending a practice that began in earnest with UNCED, nongovernmental organizations were permitted to represent their views to the CSD; about 1,400 of them were accredited to make representations.

The General Assembly also created two entities to work alongside the CSD: the Inter-Agency Committee on Sustainable Development (IACSD) and the Department for Policy Coordination and Sustainable Development. The former organization's task is to improve sustainable development coordination among UN agencies. The latter provides staff support for the CSD and IACSD and acts as a central office for promoting United Nations-wide sustainable development efforts.[67] Environmentalists feared that the CSD and these other agencies would usurp many of the powers of the UN Environment Program (UNEP) and divert attention away from UNEP's priorities (i.e., environmental protection). While this may be true, from a developmental perspective the creation of the CSD can be seen as a victory for those demanding greater consideration of international equity: environmental protection *and* economic development, not just the environment, are the goals of the CSD.

The Global Environment Facility

The Global Environment Facility (GEF) was first established in 1991 as a three-year pilot program to fund projects addressing stratospheric ozone depletion, climate change, biological diversity, and international waters. It is jointly administered by the UN Development Program (UNDP), UNEP, and the World Bank (but mostly the latter). The GEF was reestablished as an interim body during UNCED to ensure new and additional funds on grant and concessional terms to the developing countries to promote environmental programs and pay for the provisions of the agreements on limiting climate change and preserving global biodiversity. However, the GEF would continue to focus its funding on *global* environmental problems of concern to the North, rather than the local problems of most concern to the developing countries.[68] The GEF was the result of North-South compromise, like so much of UNCED. The developed countries wanted environmental funding to come from the GEF, which would remain under the aegis of institutions they dominate, namely the World Bank. The United States expressed outright opposition to the GEF almost to the last minute, fearing that it would set a precedent for funding global environmental agreements, most notably the nascent climate change convention. Only in the spring of 1992 did the United States agree to support a restructured GEF.[69]

The developing countries initially proposed a new "green fund" that would be controlled by GEF members through a one country, one vote governing system. When efforts to create a totally new fund failed, developing countries called for the GEF to be tied more closely to the United Nations system, which they saw as being more democratic and more sensitive to their development needs and economic priorities. Hence, the UNDP and UNEP continued to participate in it. Agreement by the South to make the GEF the interim financial mechanism to fund UNCED programs was made in exchange for institutional reforms that bring it closer to UN control, specifically toward the conferences of the parties of the climate change and biodiversity conventions.[70] At Rio the delegates agreed to structure the GEF to permit nongovernmental observers and to "Ensure a governance that is transparent and democratic in nature, including in terms of decision-making and operations, by guaranteeing a balanced and equitable representation of the interests of developing countries and giving due weight to the funding efforts of donor countries."[71] In addition, the GEF mandate was expanded to encompass development projects geared toward preventing land degradation, a concern of developing countries. What is more, the parties to the Biodiversity and Climate Change Conventions were themselves to control policy, program priorities, and eligibility criteria for financial transfers to help countries meet the provisions of the two conventions, duties that would otherwise be under the sole authority of the GEF and the World Bank.

In 1993 the United States and other donor countries pledged $2 billion dollars over three years to replenish the GEF and also agreed to share control of GEF by introducing a new voting system in the 32-country council. The GEF restructuring was agreed at the GEF participants meeting held in Geneva during March 1994. Under the new system, Assembly decisions require "double weighted majority" support of both sixty percent of the total number of members and sixty percent of those members who have contributed sixty percent of the funds. The governing Council consists of 16 members from developing countries, 14 from developed countries, and two from the former communist countries of Europe and the former Soviet Union. While this does not fully satisfy the developing countries, it is nonetheless unusual in that it gives recipients of aid more say in how to allocate funds donated by others, in this case the industrialized countries.[72]

There was hope, especially among delegates from developing countries, that the GEF would be used to fund national environmental priorities. But issues of vital interest to the developing countries, such as water pollution and sanitation, soil erosion and salinization, urban air quality and water shortages received much less attention at the Earth Summit than the South and its supporters in the North had hoped. Instead, it was agreed that existing aid programs would finance such programs.[73] Thus the GEF only funds projects with global significance: climate change, biodiversity, international waters, and ozone depletion. It is thus much easier for developing countries to claim that GEF funding is for "services rendered" to the rest of the world for addressing common environmental problems and, therefore, that such funding should not be part of traditional development assistance programs. Nevertheless, prospects for significant new funds in the near to medium term were not good, especially in light of the tremendous demand for development aid from the once communist countries of Eastern Europe and the former Soviet Union.[74]

The GEF is indicative of how willing—or unwilling—countries may be to turn the rhetoric of Rio into reality. While its projects focus on programs with global effects (oceans, climate change, biodiversity, and ozone), thereby neglecting the bulk of issues important to developing countries, the GEF does share burdens (developed countries contribute most of the GEF's funds), benefits (developing countries are the recipients of those funds) and decision making authority (developing countries are part of a double-majority voting system that gives recipients of aid a say in how funds are distributed). To be sure, while the amounts of money pledged to the GEF and distributed by it are significant, they are not large compared to need. The GEF experience shows how difficult it is to translate the equity provisions of international environmental institutions into concrete benefits for the environment and for the world's poor countries.

Conference Statements

The various agreements and conventions at the Earth Summit were supplemented by statements from the UN leadership, diplomats and statespersons. They were rhetorical manifestations of international equity in the context of sustainable development. For example, Maurice Strong, Secretary General of UNCED, declared that:

> Traditional notions of foreign aid and of the donor-recipient syndrome are no longer an appropriate basis for North-South relations. The world community must move towards a more objective and consistent system of effecting resource transfers similar to that used to redress imbalances and ensure equity within national societies.[75]

> The tragedy is that poverty and hunger persist in a world never better able to eliminate them. This is surely a denial of the moral and ethical basis of our civilization as well as a threat to its survival. Agenda 21 measures for the eradication of poverty and the economic enfranchisement of the poor provide the basis for a new worldwide war on poverty. Indeed, I urge [the delegates at the Earth Summit] to adopt the eradication of poverty as a central objective of the world community as we move toward the twenty-first century.[76]

Many leaders of the developed countries felt it necessary to invoke at the Earth Summit notions of international equity, justice and fairness. For example, the Prime Minister of Portugal and then President of the European Community's European Council reminded the world that the EC and its member states ranked first in the world in the provision of development assistance and that the "war on poverty must continue to be a top priority." The Chancellor of Germany said that his country was "firmly determined to live up to [its] responsibility towards the developing countries" and then committed Germany to increase its ODA to the 0.7 percent target. The Prime Minister of the Netherlands said that "people are not prepared to tolerate a situation in which others do not enjoy their rightful share in the prosperity of our world," going on to call for an eradication of world hunger and poverty. The Prime Minister of Sweden said that "it should be a moral duty for all countries of the rich world to meet the goal of providing 0.7 percent of GNP to ODA." The Prime Minister of Norway said that the "poor must be brought home from their exile of bondage and humiliation. Fair distribution of wealth and opportunity must be provided."[77]

Even U.S. President Bush, one of the participants at Rio seemingly least interested in establishing rights to economic justice on a global scale, declared his country's commitment to increase aid to the poorer countries:

> We come to Rio recognizing that the developing countries must play a role in protecting the global environment, but will need assistance in pursuing these cleaner growths. So we stand ready to increase United States international environmental aid by 66 percent above the 1990 levels, on top of the more that $2.5 billion we provide through the world's development banks for Agenda 21 projects.[78]

According to Philip Shabecoff, a journalist and close observer of the entire UNCED process who was permitted extraordinary access to the official UNCED negotiations:

> Virtually every speaker, North or South, from every continent, acknowledged that his or her country was in the same leaking lifeboat with every other country and only cooperation would save them. In the preparatory process, the industrialized nations had been concerned primarily with the environment, the poorer countries with economic development. It was almost as if they were preparing for two different conferences. At the summit the two goals merged, at least in the words of the national leaders.[79]

Statements by heads of state and government at the Earth Summit must be taken with many of the proverbial grains of salt, to be sure. But one must still ask why so many leaders from the developed North found it necessary to invoke notions of international beneficence and conceptions of a world in which the wealthy ought to help the poor.

Earth Summit II and III

In 1997 the United Nations held an UNCED follow up conference. As during the Earth Summit—if not even more so—the "Rio+5" conference was infused with considerations of international equity. It was generally agreed by delegates at "Earth Summit II" that UNCED was successful in giving new energy to the goals of the 1972 Stockholm conference, and that it provided important guidance for environmental programs and stimulated ongoing negotiations on specific important issues like climate change. But new action on protecting the global environment was incremental, and the commitment by developed countries to provide new aid to the developing world was not adequately fulfilled. Diplomats and environmentalists realized that environmental protection would be a slow, ongoing process, as would implementation of the equity provisions of the first Earth Summit. It is probably safe to say that most developing country diplomats, and many of their developed country partners, left the meeting disappointed.[80] The UNCED ten-year follow up conference ("Rio+10") will be held in Johannesburg, South Africa. Reflecting the degree to which development has permeated international environmental deliberations, it will be called the World Conference on Sustainable Development.[81] There is no doubt, however, that its conclusions will be similar to those of the Rio+5 conference: There has been incremental movement toward achieving UNCED's equity goals, but not nearly enough has been done.

Philosophical Resemblances of the UNCED Equity Provisions

Provisions for international equity are found throughout the Rio Declaration, Agenda 21, the conventions signed at the Earth Summit (FCCC and Biodiversity Convention), and other products of UNCED. Indeed, most of the documents signed at Rio are permeated with provisions for international equity, as are the follow-on deliberations regarding implementation of UNCED agreements. These equity provisions of UNCED resemble in many ways equity arguments made by philosophers outlined in the previous chapter. While the provisions might not survive the careful scrutiny of those philosophers, they do resemble arguments made on ethical grounds, including arguments based on causality and responsibility, Kantianism, Rawlsian ethics, and other philosophies.[82]

For example, the UNCED documents resemble arguments made by some philosophers who argue that those who have caused harm to others have a responsibility to stop doing that harm, and to rectify it.[83] The Rio Declaration acknowledges that countries have "common but differentiated" responsibilities and recognizes that, while it will require comprehensive cooperation to address environmental changes, the developed countries are inordinately responsible for present global environmental conditions. Principle VII of the Declaration, which states that the "developed countries acknowledge the responsibility that they bear in the international pursuit of sustainable development in view of the pressures their societies place on the global environment and of the technologies and financial resources they command," also clearly expresses an agreement that the developed countries are inordinately responsible for historical pollution and its contemporaneous consequences, and also affirms that they also ought to act because they have the financial and other resources to do so, whereas the developing countries do not. Similar provisions are found in the Climate

Change Convention, which notes that most current and historical emissions of greenhouse gases have originated in the North and that the developed countries will take the lead in trying to reduce emissions of greenhouse gases.[84]

Many provisions in the UNCED documents resemble Kantian-like notions of equity which might suggest that countries have the responsibility to help other countries reach a stage where they can be relatively free agents (insofar as such a thing is possible in the current interdependent international system), or which might obligate countries not to impose conditions on others that they would not choose freely of their own volition. The Rio Declaration's Principle II requires countries to ensure that what they do within their own borders does not harm other countries, certainly something that those other countries would choose if they had a choice in the matter. Agenda 21 declares that world poverty is the shared responsibility of all countries, including the countries where poverty is not prevalent.[85] This effectively would help poor countries make the move toward becoming independent agents. Similarly, the FCCC requires industrialized countries to help the developing countries by providing finance and technology to meet treaty objectives, and declares that actions to address climate change should give priority to the needs of developing countries for sustained economic growth and poverty eradication. The climate treaty also declares that developing countries' effective implementation of the treaty will depend on whether the developed countries fulfill their obligations to provide aid to the South, with the understanding that the Southern states must first eliminate poverty and develop economically and socially before they can take on greater responsibilities regarding climate change.[86] Other paragraphs of the Climate Change Convention state that participants in the treaty are to fully consider the special needs of developing and least developed countries and provide them with aid so that they might in future meet treaty provisions.[87] Similar provisions are contained in the Biodiversity Convention.[88]

Numerous provisions in the UNCED agreements and conventions resemble Rawlsian arguments, which emphasize actions that benefit the least advantaged members of a community and which do so in ways that provide for equality of opportunity.[89] Resembling conceptions of equity that focus on the least advantaged, Principle VI of the Rio Declaration emphasizes the "special situation and needs of the developing countries, particularly the least developed and those most environmentally vulnerable," pointing out that their needs shall be given "special priority." Agenda 21 is permeated with references to the need for substantial special help (e.g., new funds and concessional technology transfers) to the developing countries, especially the least developed countries, to help them develop in such a way that simultaneously leads to their economic progress and environmental protection.[90] Similarly, the statement of forest principles calls for the "eradication of poverty"[91] as well as "new and additional resources"[92] and "access to and transfer of environmentally sound technologies and corresponding know-how on favorable terms, including concessional and preferential terms,"[93] for the least advantaged countries of the world. The FCCC declares that special consideration should be given to the "specific needs and special circumstances" of developing countries.[94] The Biodiversity Convention has similar provisions for meeting the special needs of the developing countries. For example, Parties to the Biodiversity Convention are to take full account of the "specific needs and special situation of least developed countries" when determining requirements for new funding and technology transfer.[95]

Resembling Rawls's equal opportunity principle, the climate and biodiversity conventions ensure that the poor countries will be able to participate in the governance of the treaties.[96] The Climate Change Convention indicates that its financing body must have "equitable and balanced representation" of parties to the convention "within a transparent system of governance."[97] The Biodiversity Convention funding mechanism is to operate "under the authority and guidance of, and be accountable to, the Conference of the Parties" and shall operate "within a democratic and transparent system of governance."[98] In a similar vein, the GEF was restructured to ensure that it operated in a democratic manner and guaranteed the representation of the developing countries,[99] and the Commission on Sustainable Development was designed to have equitable representation that ensures the participation of the developing countries.

In short, many of the provisions for international equity in the UNCED agreements, conventions and follow-on negotiations can be viewed as codifications of equity concepts that abound in the philosophical literature—even though the philosophers were rarely explicitly invoked in the UNCED deliberations and documents. This description of philosophical resonance is offered as further support, if such support is needed, for the contention that considerations of equity were indeed part of the UNCED deliberations and the agreements and conventions that emanated from them.

Conclusion

A reading of the UNCED agreements, conventions and deliberations prompts the hypothesis that what happened at Rio was more than rhetorical wrapping for self-interested state policies, and that equity considerations were included in part because they may lead to a fair and just distribution of global benefits and burdens (not just because they made the signing governments look noble to their own citizens and to citizens and statespersons in other countries). The persuasiveness of equity provisions throughout UNCED statements, agreements and conventions, and the extensive negotiation that went into their inclusion are unusual, exceeding previous efforts to incorporate equity into international relations, most notably the New International Economic Order and UN Law of the Sea Convention. If these provisions are taken seriously, this would be an important and potentially very significant change. International environmental deliberations, most notably UNCED, could have real impact in promoting international equity, at least in the environmental arena, with implications for poor countries and people worldwide. The inclusion of equity considerations in the UNCED agreements suggests that we may be experiencing the earliest stages of a modest shift in worldwide consciousness about obligations of rich countries to poor countries. However, questions remain: How and why did international equity emerge at the end of the last century to become a central part of international environmental deliberations, and why did the United States, particularly during the Clinton administration, begin to join this emerging consensus? These are central questions that subsequent chapters begin to answer.

Notes

1   The Rio Declaration (A/Conf.151/5/Rev.1, 13 June 1992) is reprinted in United Nations, *Report of the United Nations Conference on Environment Development, Vol I: Resolutions Adopted by the Conference* (New York: United Nations Publications, 1993) [UNCED A/CONF.151/26/Rev.1 (Vol. I)], pp. 3-8. Even the title of the Declaration—the Rio Declaration on Environment *and* Development—was a concession to the developing countries. The United States and some other developed country governments wanted it to be called the "Earth Charter," suggesting that the primary goal of the conference would be environmental protection and de-emphasizing the developmental goals of UNCED. See "Prepared for Rio," *Environmental Policy and Law* 22, 2 (1992), pp. 129-30.

2   Principle 1.

3   Principle 25.

4   Principle 6.

5   Principle 7.

6   Para. 1.4. Agenda 21 can be found in United Nations, *Report of the United Nations Conference on Environment and Development, Volume I*, pp. 9-479. Agenda 21 does

not contain any mandatory requirements that states must follow. It is a "soft law" agreement that exhorts countries to be bound by its commitments.

7   Para. 3.2.

8   Para. 33.10. Chapter 33 addresses financial resources and mechanisms. It was the most controversial part of the UNCED negotiations and was agreed only in the eleventh hour of the Rio Summit. Chapter 33 discussions (and as a result discussions regarding the financial sections of all other chapters) were held in completely closed-door sessions.

9   Para. 33.13. At the Earth Summit the U.S. refused to be bound by the 0.7 percent goal. Indeed, the latter portion of this paragraph ("to the extent that they have not yet achieved that target. . .") was intended to cover only the U.S. and Switzerland. (*Earth Summit Bulletin* 2, 9 [June 1992)] and 2, 13 [June 1992].) The Scandinavian countries favored a commitment by all OECD countries to meet the target of 0.7 percent of GNP before 2000. Other OECD countries refused to set a date for reaching that target, although they committed themselves to it. The wording of this paragraph suited the United States, despite its objections to setting specific commitments to funding, because it has never agreed to the 0.7 percent target. To "reaffirm" its commitments required the U.S. to do nothing at all.

10   Para. 33.16(a)(iii). The last point, like so much of Agenda 21 and the other UNCED agreements, is a clear compromise between North and South. The South got agreement that there should be new funds without new forms of conditionality. The North ensured that funding would be on mutually agreed terms (i.e., subject to their prior agreement).

11   Para. 34.14(b). The U.S. government took exception to the thrust of this provision.

12   These figures are estimates at best. While the amount required to fund Agenda 21 was the result of calculating the costs of the many proposed programs, one member of the secretariat described how the cost of one major program was arrived at this way: "We were at a party and drinking a lot of wine, when somebody said, 'What about this number?' And we all said, 'Yeah, let's take that number.'" Philip Shabecoff, *A New Name for Peace: International Environmentalism, Sustainable Development, and Democracy* (Hanover, NH: University Press of New England, 1996), p. 146.

13   Paul Lewis, "Environmental Aid for Poor Nations Agreed at the UN," *New York Times*, 5 April 1992. Money pledged at the Earth Summit for sustainable development was estimated at between $3 and $6 billion, but how much of this was new money is unclear. Determining total pledges and the extent to which they are really "new and additional" is nearly impossible even for UN insiders. Barbara Ruis, Associate Expert (Legal), Human Development, Institutions and Technology Branch, Division for Sustainable Development, United Nations Department for Policy Coordination and Sustainable Development, telephone interview by author, 14 December 1995.

14   See Authoritative Statement of Forest Principles in United Nations, *Report of the United Nations Conference on Environment and Development, Volume I*, pp. 480-85.

15   Quoted in Stanley P. Johnson, *The Earth Summit* (London: Graham and Trotman, 1993), p. 103.

16   Editorial, *Environmental Policy and Law* 22, 4 (August 1992), p. 202. The U.S. government's principle objective at Rio, according to Environmental Protection Agency administrator William Reilly, was to preserve as much as possible of the world's forests (because they absorb carbon dioxide). "Rio Conference on Environment and Development," *Environmental Policy and Law* 22, 4 (August 1992), p. 223.

17 Editorial, *Environmental Policy and Law* 22, 4 (August 1992), p. 201.

18 Johnson at p. 110 describes the culmination of efforts to negotiate a forest treaty this way: "Over a period of two years, the West's finest diplomatic brains were totally outmaneuvered and the fact that these diplomatic efforts were reinforced by political commitments at the highest level (from President Bush, for example, or from Chancellor Kohl of Germany) appeared to count for nothing. . . . Some have argued that Malaysia and India would have been forced to modify their hostility to the very idea of opening negotiations on a forest treaty if the West had given a more substantial earnest [*sic*] of its good intentions on the general issue of resources and technology, instead of waiting until the very last minute to haggle over finances." Both environmentalists and forest industry lobbyists regarded the forest principles as "so toothless as to be meaningless." Shabecoff, *A New Name for Peace*, p. 164.

19 Scott A. Hajost, "The Role of the United States" in Luigi Campiglio et al., eds., *The Environment After Rio: International Law and Economics* (London: Graham and Trotman, 1994), p. 18.

20 Para. 7(a).

21 Para. 10.

22 Para. 11.

23 In September 2000 the UN's Economic and Social Council established the Forum on Forests, a subsidiary body intended to build on the Rio Forest Principles and chapter 11 of Agenda 21.

24 Over-consumption in the developed countries is one of the foundations of the developing countries' justifications for demands for more serious consideration of North-South equity. See, for example, Luis N. Camacho, "Consumption as a Theme in the North-South Dialogue," *Philosophy and Public Policy* 15, 4 (Fall 1995), pp. 32-34.

25 United Nations, *Energy Statistics Yearbook* (New York: United Nations Publications, 1994).

26 The Framework Convention on Climate Change is reprinted at 31 *International Legal Materials* 849 (1992).

27 The U.S. signed the treaty, but it insisted that the convention document not set specific targets for reducing carbon dioxide emissions nor obligate countries to take specific actions to reduce emissions of greenhouse gases. The U.S. Senate ratified the treaty in October 1992, making the U.S. the first industrialized country (and third country overall) to do so. In mid-1996, the Clinton administration expressed its support for negotiation of a legally binding protocol to limit greenhouse gas emissions.

28 "Climate Change Treaty Comes into Force," *International Environment Reporter*, 23 March 1994.

29 Art. 4, para. 2(a).

30 This is based on discussions with a group of Northern delegates and their deputies at the United Nations Intergovernmental Negotiating Committee on Climate Change (INC) meeting, 9-10 February 1995, including Rafe Pomerance, U.S. Deputy Assistant Secretary for Environment and Development and Sven Aass, Norwegian delegate to the United Nations.

31 Art. 3, para. 1.

32 Art. 3, para. 2.

33 Art. 3, para. 4.

34 Art. 3, para. 5.

35 See Art. 12.

36 Art. 4, para. 3.

37 Art. 4, para. 5.
38 Art. 4, para. 7.
39 Art. 4, paras. 8-10.
40 Richard Mott, "The GEF and the Conventions on Climate Change and Biological Diversity," *International Environmental Affairs* 5, 4 (Fall 1993), p. 302.
41 Art. 11, para. 1.
42 Ibid.
43 Art. 11, para. 2.
44 Art. 21, para. 3.
45 The Biodiversity Convention is reprinted at 31 *International Legal Materials* 818 (1992).
46 Pratap Chatterjee and Mathias Finger, *The Earth Brokers: Power, Politics and World Development* (London: Routledge, 1994), p. 43.
47 Alan E. Boyle, "The Convention on Biological Diversity," in Luigi Campiglio, et al., eds., *The Environment After Rio: International Law and Economics* (London: Graham and Trotman, 1994), pp. 113-14.
48 Art. 1. The Convention thus reflects both the differing interests of North and South. The North wanted to protect biodiversity and the sustainable use of its components. Alternatively, the South wanted fair and equitable sharing of the benefits derived from the use of its genetic resources. Thus, according to Boyle, a "trade-off between conservation and economic equity is at the heart of the Convention and makes it unusual among environmental agreements." Boyle, p. 116.
49 Art. 15, para. 7.
50 Art. 16, para. 3.
51 Art. 16, para. 2. The Bush administration's refusal to sign the treaty was mostly a result of its opposition to the technology transfer provisions of Articles 15 and 16. Matthew Townsend, "The International Transfer of Technology," *Environmental Policy and Law* 23, 2 (1993), pp. 66-69.
52 Art. 19, para. 2.
53 Art. 20, para. 2.
54 Art. 20, para. 4.
55 Art. 20, para. 5.
56 Quotes are from Art. 21, para. 1. Read in conjunction with the paragraph requiring funding to meet the incremental costs incurred by the developing countries, paragraphs on the funding mechanism may be interpreted as requiring a "blank check" from the developed countries. Hence statements for the record by 19 countries made during signature of the Final Act declare that those countries' interpretations do not require them to fulfill demands made by the funding mechanism. Wolfgang E. Burhenne, "Biodiversity-Legal Aspects," *Environmental Policy and Law* 22, 5/6 (1992), p. 325.
57 GEF grants go to countries with annual per capita incomes of under $4000. Its funds are focused, however, on environmental programs that have global benefits (climate change, ozone, biodiversity, oceans). Thus programs dear to the poor countries—sanitation, local irrigation, etc.—are generally not eligible for funding from the GEF.
58 Mott, "The GEF," p. 308.
59 Ibid.
60 Ibid., p. 309.
61 Boyle, p. 124. Similar provisions for "additionality" are found in other UNCED documents. Such provisions are absent from agreements as recent as the 1987

Montreal Protocol. A precedent was set, however, in the London amendments to the Montreal Protocol, which do call for additional funds to help developing countries reduce their emissions of ozone-destroying chemicals.

62 Burhenne, p. 325.

63 The treaty acknowledges the developing countries' "sovereign right to exploit their own resources pursuant to their own environmental policies" in Art. 3, a verbatim restatement of Principle 21 of the 1972 Stockholm Declaration on the Human Environment.

64 See United Nations, General Assembly, Resolution 47/191, "Institutional Arrangements to Follow Up on the United Nations Conference on Environment and Development" (1993), A/C.2/47/l.61, reprinted in *Environmental Policy and Law* 23, 1 (1993), pp. 41-43. See also "Assembly Creates Sustainable Development Commission, Endorses 'Agenda 21,'" *UN Chronicle* 30, 1 (March 1993), pp. 80-81, and "From Words to Deeds: New Commission Builds on Earth Summit Legacy," *UN Chronicle* 30, 3 (September 1993), pp. 62-63.

65 A summary and assessment of the CSD's evolution before the 1997 meeting can be found in "Summary of the Fourth Session of the UN Commission on Sustainable Development," *Earth Negotiations Bulletin* 5, 57 (6 May 1996).

66 Resolution 47/191, Para. 6, recommends that the CSD "consist of representatives of 53 States from among the Member States of the United Nations and its specialized agencies for three-year terms, with due regard to equitable geographical distribution," and suggests that the regional allocation of seats could be the same as in the UN Commission on Science and Technology for Development.

67 These and related developments are described in the *Yearbook of International Environmental Law 1992* (Boston: Graham and Trotman, 1993), passim.

68 See Richard N. Mott, "Financial Transfers Under the Conventions on Climate Change and Biodiversity," *International Environmental Affairs* 7, 1 (Winter 1995), p. 71.

69 The change in U.S. policy, as explained in chapter 7, can be attributed, at least in part, to the departure of influential White House chief of staff John Sununu. The Bush administration certainly preferred the GEF to a separate "green fund," which the developing countries had been demanding.

70 Mott, "The GEF," p. 300.

71 Agenda 21, Ch. 33.13(a)(iii).

72 See Paul Lewis, "Rich Nations Plan $2 Billion for Environment," *New York Times*, 17 March 1994. The restructuring and specific responsibilities of the Assembly, Council and Secretariat are described in Global Environment Facility, "Instrument for the Establishment of the Restructured Global Environment Facility," Report of the GEF Participants Meeting, Geneva, 14-16 March 1994 (publication date 31 March 1994), reprinted in *Environmental Policy and Law* 24, 4 (1994), pp. 192-200.

73 Nurul Islam, "An UNCED Overview," *International Environmental Affairs* 5, 3 (Summer 1993), p. 184.

74 See Hilary F. French, *Partnership for the Planet: An Environmental Agenda for the United Nations*, Worldwatch Paper 126 (Washington: Worldwatch Institute, July 1995), p. 25. For recent details of GEF funding allocations, see the GEF Web site at: <http://www.gefweb.org>.

75 United Nations, *Report of the United Nations Conference on Environment and Development, Vol II: Proceedings of the Conference* (New York: UN Publications, 1993) [UNCED A/CONF.151/26/Rev.1 (Vol. II)], p. 47.

76   United Nations, *Report of the United Nations Conference on Environment and Development, Vol II*, p. 48.

77   United Nations, *Report of the United Nations Conference on Environment and Development, Vol III: Statements Made by Heads of State or Government at the Summit Segment of the Conference* (New York: UN Publications, 1993) [UNCED A/CONF.151.26/Rev.1 (Vol. III)], quotes from pp. 6, 29, 105, 156, and 191 respectively.

78   Ibid., p. 78. Bush's statement is especially noteworthy—despite its very limited commitments—because not long before the conference the U.S. government had refused to acknowledge the need for new resources to help developing countries protect their environments.

79   Shabecoff, *A New Name for Peace*, p. 165.

80   Derek Osborn and Tom Bigg, *Earth Summit II: Outcomes and Analysis* (London: Earthscan Publications, 1998).

81   See United Nations, "Rio+10: The World Summit on Sustainable Development," <http://www.un.org/rio+10/>.

82   The objective of this section is to show how many of the provisions of the UNCED agreements and conventions *resemble* equity arguments made by philosophers. It is hoped that this will further establish that these provisions are indeed about "equity." The goal here is not to establish any particular ethical or philosophical meaning in the UNCED provisions, nor to justify those provisions on these grounds. Nor does this section make any claim to giving these philosophical arguments more than the most cursory treatment. Philosophers debate the meaning and merits of these and many other approaches to equity; this study does not attempt to join that debate. For an examination of equity principles as they might apply to global environmental issues, see Matthew Paterson, "International Justice and Global Warming," in Barry Holden, ed., *The Ethical Dimensions of Global Change* (London: Macmillan, 1996), pp. 181-201. See also Chris Brown (from whom Paterson derives many of his underlying arguments), *International Relations Theory: New Normative Approaches* (New York: Columbia University Press, 1992) and the works cited therein.

83   See Henry Shue, "Equity in an International Agreement on Climate Change," in Richard Samson Odingo et al., eds., *Equity and Social Considerations Related to Climate Change* (Nairobi: ICIPE Science Press, 1995), pp. 385-92, and Shue, "Subsistence Emissions and Luxury Emissions," *Law and Policy* 15, 1 (January 1993), pp. 39-59. For more general arguments with regard to causality and responsibility, see Brown.

84   Art. 4, para. 2(a).

85   Para. 3.2.

86   Art. 4, para. 7.

87   Art. 4, paras. 8-10; Art. 11, para. 1.

88   Art. 20, para. 4.

89   Compare the general arguments in John Rawls, *A Theory of Justice* (Cambridge, MA: Harvard University Press, 1971) and Charles Beitz, *Political Theory and International Relations* (Princeton: Princeton University Press, 1979).

90   See especially Agenda 21, chapters 1-3, 33, and 34.

91   Para. 7(a).

92   Para. 10.

93   Para. 11.

94   Art. 3, para. 2.

95   Art. 20, para. 5.
96   Rawls's principle of fair opportunity says that offices and positions ought to be open to all under conditions of fair equality of opportunity. Rawls, p. 302.
97   Art. 11, para. 2.
98   Art. 21, para. 1.
99   Agenda 21, Ch. 33.13(a)(iii) describes the types of changes to the GEF that were expected. See above for a description of the revised GEF structure.

4   A History of International
Equity in Global
Environmental Politics

The international equity provisions of UNCED statements and agreements
did not appear suddenly. They are the outgrowth of international
environmental deliberations over the last few decades. This chapter
summarizes that evolution by looking at selected international
environmental negotiations and instruments that have significant provisions
for international equity. This summary provides the important historical
context for the main arguments of this book, introduced in previous
chapters and expanded in subsequent ones, which focus specifically on the
U.S. posture toward international environmental equity in the context of
UNCED and its follow-on agreements.

Many developed countries, especially the United States, have
traditionally sought to prevent the setting of precedents that recognize the
demands of developing countries for greater international equity. This was
the case during most phases of NIEO and the Law of the Sea negotiations,
as well as during less comprehensive undertakings, such as international
negotiations on the broadcast spectrum.[1] However, such traditional
developed country policy, while still evident, was muted throughout the
1990s. The London amendments to the Montreal Protocol, UNCED, and
international environmental conventions signed in the 1990s were
permeated with discussions about, and provisions for, international
environmental equity. This emergence of international equity as a
prominent part of international environmental negotiations and agreements
is unusual by historical standards of international relations.

International equity has been on the agenda of international politics
for decades. Demands by poor countries for more equitable treatment in
international economics were evident when the United Nations Charter was
being negotiated. This was reflected, for example, in the placement of the
Economic and Social Council alongside (at least nominally) the other
principal organs of the United Nations.[2] North-South relations came onto
the international agenda with much greater prominence in the 1960s as a
result of post-war decolonization.[3] By the mid-1960s, the developing
countries acquired a majority of votes in the UN General Assembly, putting

them in a stronger position to push their economic demands. This new influence was reflected in the UN Conference on Trade and Development (UNCTAD), set up in 1964, and the establishment of the Group of 77 "non-aligned" developing countries (G-77, now more than 130 countries). International equity in environmental agreements "went global" with the 1972 UN Conference on the Human Environment (see below), but the developing countries' newfound voting power in the United Nations proved to be of limited utility. Their demands did not initially lead to substantial new initiatives for international equity. Nevertheless, equity has slowly been gaining prominence—in fits and starts—in international environmental relations ever since. Many of the demands reflected in the NIEO were, to varying degrees, codified in the Law of the Sea, the Montreal Protocol amendments, and the documents that established UNCED—as well as the UNCED agreements and conventions described in the previous chapter.

The New International Economic Order

The "New International Economic Order," which refers to the package of demands made by developing countries in the 1960s and debated in a 1974 special session of the UN General Assembly, was an effort by the those countries to restructure the world economy in a manner that would redistribute global financial resources to the South.[4] The NIEO declarations called for: (1) trade reforms, including the removal of barriers to trade and stabilization of markets for basic commodity exports from the developing countries, in order to reduce the disparities in the prices of basic commodities from the South and manufactured goods from the North and to increase the market share of Southern exports; (2) monetary reforms, including the stabilization of inflation and exchange rates, increased funding from the IMF and the World Bank, and greater participation by the developing countries in decision making in these and other international financial institutions; (3) assistance from the North to promote technology transfer and increase industrial production in the South;[5] (4) economic sovereignty, whereby the developing countries would control their own resources and be free to regulate the activities of multinational corporations; and (5) increased foreign aid from North to South, at minimum 0.7 percent of the donor countries' GNPs.[6]

In the wake of the energy crises of the 1970s—which enhanced the bargaining leverage of the developing countries—the European Community agreed to some of the NIEO demands (in the Lomé Convention[7]), and even the United States made minimal concessions to them.[8] Nevertheless, for the most part the developed countries were not receptive to NIEO demands, at least not those that would require substantial intervention in the global economy to redistribute income toward the poor countries. The developed countries, believing that intervention in the global marketplace would do more harm than good for the world economy and for the developing countries, pushed instead for increasing reliance on free-market forces, as evidenced by ongoing negotiations under the auspices of GATT. This push was strongly supported in the 1980s by the U.S. government under Ronald Reagan. The Reagan administration made promotion of a global *laissez faire* market its primary international objective after containing communism and the Soviet Union.

The NIEO met with very limited success.[9] In most respects the South gave up its most passionate efforts to transform the world economy to its advantage and instead began to gradually focus on meeting its demands in specific issue areas. The debate surrounding the 1972 UN Conference on the Human Environment (UNCHE) held in Stockholm, Sweden, occurred in the midst of these nascent but mostly ineffective calls by the South for greater international equity.

The Stockholm Conference

One hundred and fourteen countries participated in the Stockholm Conference.[10] Before Stockholm, most international environmental agreements focused on scientific issues. The Stockholm Conference addressed broader political, social, and economic issues as well. North-South economic differences were a major element of the preparatory meetings before the conference and of the conference itself. As was the case to an even greater extent twenty years later in Rio, Stockholm was a soapbox for opposing perspectives: the North's, emphasizing humankind's adverse impact on the environment, and the South's, which focused on economic and social development. Participants from the South generally laid the blame for much of the poverty and pollution in developing countries on practices by the North that exploited the poor countries. Developing countries feared that the Stockholm agreements might have adverse effects on their own development. They worried that stricter environmental standards in the developed countries would raise the price of manufactured products necessary for development in the South, further exacerbating already unfavorable terms of trade. They also worried that scarce development funds would be diverted away from economic development to strictly environment-related projects. These differences were bridged, at least in dialogue, by advancement of the notion that protection of the environment is an integral component of effective socioeconomic development.[11]

At Stockholm developing countries demanded sovereignty over their biological resources, technology transfers from North to South, and access to additional financial resources. The most divisive topic of deliberation was the demand by developing countries that they share in the fruits of biotechnology derived from their biological resources.[12] The South began to connect its demands for technology transfer to biological diversity by suggesting that access to genetic material could be made a function of concessional access to technology. The industrialized countries were persuaded to include items of concern to the South in the conference declarations. These items included provisions addressing improvements in shelter, food, and access to clean drinking water, along with other broader development concerns of the South. The Conference also agreed that all states should have sovereignty over their biological resources. Without developing country efforts, the Stockholm conference would have focused almost entirely on issues like pollution, population growth, resource conservation, limits to growth, and the like.[13]

Considerations of international equity appeared in the Stockholm Declaration.[14] Paragraph 4 declared that environmental problems in the developing countries are caused primarily by underdevelopment, that millions of people there live "far below minimum levels required for a decent human existence, deprived of adequate food and clothing, shelter and education, health and sanitation," and therefore "the industrialized countries should make efforts to reduce the gap between themselves and the developing countries." Principle 9 stated that "Environmental deficiencies generated by the conditions of underdevelopment and natural disasters pose grave problems and can best be remedied by accelerated development through the transfer of substantial quantities of financial and technological assistance. . . ." Principle 12 called on the industrialized countries to take into account the particular requirements of the developing countries and "any costs which may emanate from their incorporating environmental safeguards into their development planning and the need for making available to them, upon their request, additional international technical and financial assistance for this purpose."[15] Principle 21 declared that countries have the "sovereign right" to exploit their own resources as they choose, and the "responsibility to ensure that activities within their jurisdiction or control do not cause damage to the environment" of other countries.

These calls for new funding, concessional technology transfers, recognition of the "special burden" that environmental protection could pose for the developing countries, and other statements were important rhetorical steps toward greater international equity. The decision taken at Stockholm to have the headquarters of the new United Nations Environment Program in Nairobi, rather than Geneva or New York—or even Stockholm—was an important symbol of the growing awareness of the need to accommodate the demands of the developing countries in the environmental field. In sum, the Stockholm conference contributed to a greater *awareness* of international equity issues, especially as they relate to the environment, but beyond this little of substance to actually promote international equity emerged from Stockholm.

The Law of the Sea

The Law of the Sea treaty (LOS), signed in 1982 by almost 160 countries after more than a decade of complicated negotiations, incorporated several provisions for international equity.[16] It entered force in 1994 after the required 60 ratifications. The Third UN Conference on the Law of the Sea was initiated in 1973 by the United States and the Soviet Union to protect freedom of navigation at a time when developing coastal states were claiming larger areas of adjacent waters as their national jurisdictions. The treaty was negotiated during the heyday of the NIEO, which made developed countries wary of Southern proposals for collective ownership of the deep seabed, global taxes, and technological and financial transfers. But developed maritime countries were persuaded by self-interest to repress many of their misgivings and accept several of South's demands for the equity provisions in the treaty.

Demands by the South that parts of the oceans be declared the "common heritage of mankind" had to be taken seriously if maritime powers were to ensure predictable access to coastal waters and straits of passage that could be restricted if adjacent developing coastal states declared those areas sovereign territorial waters. Northern shippers and fishers, as well as the navies of the maritime military powers—most notably the United States—wanted unrestricted access to straits of passage.[17] Thus, in the 1970s the United States and other developed maritime powers essentially traded transit passage through international straits and freedom of navigation on the high seas for the developing countries' demands for international regulation and control of deep seabed mining—including the concept of "common heritage of mankind"—and extended exclusive economic zones.[18] Conceivably the United States and other maritime powers could have chosen to use military force to guarantee rights of passage, but they did not do so, instead agreeing to bargain their demands for access with those of the developing countries for equity.

This was a situation in which the developing countries—at least the many coastal states among them—had newfound bargaining leverage that pushed the North to take Southern demands seriously. The developed countries were faced with a clash between their national objectives and the unique characteristics of maritime "commons." Like other global commons—such as the ozone layer, geosyncronous orbit, and the atmosphere—the seas could benefit all users if managed cooperatively. But if a few countries chose not to cooperate because they believed that doing so would promote their own short-term objectives, many states—in this case the maritime developed countries—might suffer. The Northern countries could not achieve their objectives without the participation of the South. The poor and weak developing countries were therefore able to extract concessions from a reluctant North. Thus, according to some observers, the Law of the Sea is the closest the South has come to constructing its "ideal" regime because the treaty "gave developing countries ready access to decision-making forums and invested them with more influence and power than they could ever have claimed on the basis of their national power capabilities."[19]

Provisions of the Law of the Sea provided for equitable sharing of mineral resources on the seabed outside the maritime boundaries of coastal states. The deep ocean floor beyond the exclusive economic zones of coastal states was declared the "common heritage of mankind."[20] These areas were to be managed by the International Seabed Authority (ISA) and the associated Enterprise. Developing countries hoped the ISA's exclusive rights to the deep seabed would produce new financial resources to promote their own development. Final agreement divided rights to deep seabed exploitation between the ISA and private or public corporations. Corporations were required to transfer technology to the Enterprise or groups of developing countries applying for contracts "on fair and reasonable terms."[21] They were also required to share mining sites with the Enterprise. If developing countries were adversely affected, they could apply to the convention's assembly for compensation.

The Reagan and Bush administrations refused to sign what they viewed as a fatally flawed Convention on the Law of the Sea, citing especially the deep seabed provisions that they believed would prejudice U.S. corporations.[22] However, the U.S. government under President Clinton signed the Law of the Sea Treaty after negotiation of a set of interpretations that addressed the U.S.'s long-standing grievances, most notably regarding technology transfer and the allocation and funding of deep seabed exploitation rights. The changes included eliminating deep seabed mining production quotas, de-facto veto power for the developed countries in the ISA, dropping a mandatory exploration surcharge, and an end to requirements for mandatory technology transfers to help developing countries' mining programs.[23]

The Law of the Sea has produced few economic benefits from the deep seabed for poor countries, in large part because exploitation of the deep seabed is not yet economically fruitful and technology to take advantage of its resources remains in the hands of a few developed countries, including the United States. Nevertheless, while the Law of the Sea was "fixed" in accordance with U.S. wishes by the Clinton administration, its "'normative' dimension [was not] dramatically altered, and concepts such as that of greater access to shared resources and the special rights of the coastal States within their Exclusive Economic Zones (EEZs) were protected and recognized."[24]

Ozone Depletion: The Vienna Convention and Montreal Protocol

Multilateral efforts to protect the stratospheric ozone layer from depletion began with the 1985 Vienna Convention for the Protection of the Ozone Layer and the 1987 Montreal Protocol on Substances that Deplete the Ozone Layer. The Vienna Convention focused on agreement to cooperate in the gathering of information. Initial negotiations on stratospheric ozone depletion were relatively easy because it was believed that comprehensive participation was not essential for an effective agreement. Thus, negotiations for the Vienna Convention could be successful with participation of only the countries that were major produces of ozone-destroying chemicals. The South was not a significant player at the Vienna Convention.

Developing countries took a greater interest in negotiations for the 1987 Montreal Protocol, an agreement to limit the production of chlorofluorocarbons (CFCs) and other ozone-destroying chemicals. The developing countries feared that the agreement might limit their access to CFCs, which were used in ever-greater quantities in the South, primarily for refrigeration and air cooling. In the event of restrictions on CFC production, the developing countries wanted either free or highly concessional access to substitute chemicals or financial assistance to help them buy those substitutes. At least in the early stages of negotiations, most Northern countries were highly resistant to such demands. The United

States was especially determined to avoid setting up a funding arrangement that might act as a precedent for nascent negotiations on climate change.[25] But India and China were especially vocal in their insistence that they should not have to suffer from efforts to fix a problem that had resulted from actions by the industrialized countries.[26]

The 1987 Montreal Protocol did incorporate some provisions to persuade the South to join. Developing countries were permitted a modest expansion in their use of CFCs during a ten-year transitional period. They were also entitled to technology transfer to help them transition to new technologies that were not ozone destroying. However, the 1987 Protocol had no provision for a funding mechanism to help defray the costs incurred by the poorer countries transitioning to CFC substitutes. Thus the Protocol failed to elicit developing country support and many of those countries—most notably China, India and Brazil—chose not to sign the Montreal Protocol. They made their participation contingent on the creation of a fund that would provide funds in addition to prevailing aid flows to help them make the transition to CFC substitutes. By the 1989 Meeting of the Parties to the Montreal Protocol in Helsinki the industrialized countries agreed to modest measures to help developing countries acquire information, research and training, and to aid them in their efforts to garner financing for technology transfers and retooling necessary to fulfill obligations of the Montreal Protocol.[27]

Amendments to the Montreal Protocol

In light of increasing scientific knowledge and public concern in the North about ozone depletion, by the start of the 1990s there was a new emphasis on phasing out ozone-destroying chemicals rather than just limiting their production. At the second meeting of the parties at London in 1990 the developed countries agreed to substantial new efforts to bring the developing countries on board, especially those large developing countries with the most potential to derail efforts to phase out most ozone-destroying chemicals. The London amendments offered financial inducements to developing countries, including technical cooperation and a multilateral fund, financed by the North, that would help pay for their efforts to transition to new chemicals.

The United States insisted that a precedent must not be set, fearing the costs of assisting the developing countries in future efforts to address climate change. The United States wanted any fund to be administered by the World Bank, over which it had substantial control. The United States also insisted that funds come from existing development aid. But even the United States agreed—at the last minute and only after intense lobbying by other industrialized countries—that efforts to phase out ozone-destroying chemicals would be fruitless if the largest developing countries were not brought on board the agreement. According to one account, despite opposition to the fund by White House chief of staff John Sununu and director of Office of Management and Budget Richard Darman, the United States agreed to contribute to the fund after direct appeals to President Bush from EPA administrator William K. Reilly and key industry chief executive officers.[28]

Consequently, the London amendments contained several provisions for international equity that were absent from the 1987 Protocol. For example, the amendments "acknowledge that special provision is required to meet the needs of developing countries, including provision of additional financial resources and access to relevant technologies."[29] They established that developing countries have special needs and that those countries' compliance with the treaty will depend on funding and technology transfers from the more affluent parties.[30] The amendments called on parties to establish a multilateral fund "for the purposes of providing financial and technical cooperation, including transfer of technologies" to developing countries to help them comply with the treaty. Contributions to the financial mechanism "shall be additional to other financial transfers" and "shall meet all agreed incremental costs" incurred by developing countries.[31] Funds were to be provided to poor countries on a grant or concessional basis.[32] Under the amendments, the developed countries agreed to "expeditiously" transfer applicable technologies to the developing countries "under fair and most favorable conditions."[33]

Toward this end, the Interim Multilateral Fund was established in January 1991, becoming the Multilateral Fund two years later.[34] It was the first dedicated fund associated with an international environmental agreement. In its first ten years, the fund received over US$1 billion from 32 developed countries.[35] New certainty about the dangers of ozone depletion led further amendments to the Montreal Protocol for phase-outs and additional limits on more chemicals.[36] Hence the developed countries genuinely want to completely phase most out ozone-destroying chemicals worldwide. We can therefore expect the equity provisions of the amended Montreal Protocol to lead to ongoing technical cooperation and funding to help the developing countries make the transition to CFC substitutes. This will be real international environmental equity.[37]

Preparations for the Earth Summit

The United Nations Conference on Environment and Development was initiated by developed countries concerned about environmental consequences of industrialization. By the late 1980s, they had come to recognize the dangers of global environmental changes. President Bush and British Prime Minister Margaret Thatcher (the latter not previously known for her assertions of environmental awareness and concern) declared their concern for the global environment. Thatcher described environmental protection as one of the "great challenges of the late twentieth century."[38] In his election campaign Bush said that he would be the "environmental president" and that he would host an international environmental conference. In addition, at the 1989 meeting of the Group of Seven industrialized countries, the United States and its closest developed country allies declared that the global environment ought to be protected for future generations. They also recognized the links between environmental protection, economic development and poverty, declaring in the conference communiqué that protecting the global environment called for "a determined and concerted international response and for the early adoption, worldwide, of policies based on sustainable development."[39]

Not surprisingly, therefore, the earliest preparatory meetings for UNCED focused largely on the objectives of the industrialized countries.[40] However, the differences between North and South witnessed at Stockholm were also apparent: the North wanted to focus on environmental problems; the South wanted emphasis to be placed on economic development. During an UNCED preparatory meeting, Algerian *rapporteur* Ahmed Djoghlaf summarized the South's perspective: "For industrialized countries, the issue is finding ways and means of adding environmental considerations to their developed economies. But in many developing countries, development is completely missing. So how are we supposed to incorporate environment into something that does not exist?"[41] As the date for the Rio summit approached, the sentiments of the South—that an environmentally healthy planet was impossible under the prevailing circumstances of significant international inequities—became much more salient.[42]

The theme of UNCED, sustainable development, largely came from the work of the World Commission on Environment and Development (the Brundtland Commission), which grew out of the Stockholm Conference and was established by the UN General Assembly in 1983. The report of the Brundtland Commission, *Our Common Future*, emphasized the links between poverty, development and environment.[43] The report popularized the notion of sustainable development, defined by the Brundtland Commission as environmentally benign development that meets the needs of present generations without impeding future generations from meeting their own needs.[44] The Brundtland Commission was explicit in stating that the concept of sustainable development must encompass efforts to meet the essential needs of the world's poor, "to which overriding priority should be given."[45] The Brundtland Commission described the growing realization in governments and international institutions that economic development and protection of the environment were closely linked. As the report stated, "many forms of development erode the environmental resources upon which they must be based, and environmental degradation can undermine economic development. Poverty is a major cause and effect of global environmental problems. It is therefore futile to attempt to deal with environmental problems without a broader perspective that encompasses the factors underlying world poverty and international inequality."[46] The concept of sustainable development, because it brings together notions of ecology, economic development, and poverty, set the stage for incorporation of equity into international environmental discourse. It is difficult to think about sustainable development, either from the international or national perspectives, without at least implicitly thinking about equity.

The UN General Assembly established UNCED in Resolution 44/228, which was adopted at the end of 1989.[47] Echoing the report of the Brundtland Commission, the resolution called for a conference to address both environmental protection and economic and social development. The resolution was permeated—like the subsequent documents adopted at Rio—with provisions for increased international equity, especially insofar as they could contribute to environmental protection. Importantly, 44/228 served as the reference point and constitutive basis for all subsequent UNCED negotiations (especially the Rio Declaration, Agenda 21, and the Forest Principles).[48]

Numerous provisions for international equity were included in Resolution 44/228. It declares that poverty and environmental degradation are closely interrelated, that developing countries have special needs, and that the "promotion of economic growth in developing countries is essential to address problems of environmental degradation."[49] The General Assembly agreed that the developed countries bore the brunt of the responsibility for destruction of the environment: "the largest part of the current emissions of pollutants into the environment, including toxic and hazardous wastes, originates in developed countries, and therefore . . . those countries have the main responsibility for combating such pollution."[50] Resolution 44/228 calls for examination of methods for "favorable access to, and transfer of, environmentally sound technologies, in particular to developing countries, including on concessional and preferential terms [and] assured access of developing countries to environmentally sound technologies. . . ."[51] Further, the resolution declares in several places that "new and additional financial resources will have to be channeled to developing countries in order to ensure their full participation."[52] The resolution also calls on the Conference to consider the creation of a special international fund "with a view to ensuring, on a favorable basis, the most effective and expeditious transfer of environmentally sound technologies to developing countries."[53]

Resolution 44/228 was followed by multi-tiered and multi-year UNCED negotiations: one organizational meeting, four preparatory committee (prepcom) meetings, the Earth Summit, and ongoing follow-on negotiations addressing implementation.[54] The Climate Change and Biodiversity Conventions were negotiated in their own Intergovernmental Negotiating Committees (INCs), but they are generally grouped under the rubric of UNCED due to the close relationship between their objectives (in short, sustainable development vis-a-vis the atmosphere and biodiversity) and the fact that they were signed at the Earth Summit. Intergovernmental Negotiating Committee negotiations on the critical and contentious issues of institutions, financing and technology transfer also overlapped the UNCED negotiations on Agenda 21, the Forest Principles, and the Rio Declaration. The prepcoms and Earth Summit served, according to one editorial, "as a platform to discuss environmental concerns at the highest levels, and have brought poverty and development issues back onto the international agenda. In this respect, valuable papers have been produced which will always serve as invaluable references to development planners."[55]

The first meeting of the Preparatory Committee took place in August 1990 at Nairobi, Kenya, and set the terms of reference for subsequent prepcoms. At this meeting the developing countries proposed through the G-77 that there should be new financial resources for their development and technology transfers on preferential, noncommercial, and concessional terms. The United States was opposed to the transfer of technology and new resources; its position prevailed at the first prepcom and such equity issues were not part of the UNCED working group's considerations there. Prepcom II was held in Geneva in March 1991. The bulk of the many documents discussed at that meeting were provided by the UNCED Secretariat, leaving diplomats to focus on the most contentious issues. It was at this meeting that agreements later signed at Rio began to take shape. At the second prepcom the United States, the United Kingdom and France made it clear that additional assistance for general environmental programs would be unlikely.[56] The developing countries shifted their focus to linking their demands directly to the environmental objectives of the developed countries,[57] and it was at this prepcom that developed countries, including the United States, agreed to demands from the developing countries that poverty be included on the UNCED agenda as an obstacle to sustainable development.[58] In subsequent preparatory meetings and at the Rio conference the developing countries demanded that the North help them combat poverty and develop economically; otherwise, they said, it would be impossible to protect the environment. The third prepcom met in Geneva in August 1991. Substantive negotiations on Agenda 21 began at this meeting. The fourth prepcom, held in New York shortly before the Rio Earth Summit, was the setting for the most productive negotiations on the forest principles, the Rio Declaration, and the technical chapters of Agenda 21. By the conclusion of this meeting, most of Agenda 21 had been agreed by delegates.

Several areas of substantial disagreement remained at the end of the fourth prepcom, most notably paragraphs dealing with "means of implementation" that would become part of each Agenda 21 chapter, as well as sections on technology transfer, atmosphere, and forests. Questions of finance and technology transfer were so contentious that final negotiations on them had to be deferred to the Rio conference, as were questions of developed country commitments to reduce emissions of greenhouse gases under the Climate Change Convention. The Earth Summit itself became the most important negotiating session in the entire UNCED process. It consisted of Main Committee meetings (sometimes referred to as "Prepcom V"); the Plenary meeting (i.e., ministerial level statements); and several contact groups, set up by the main committee chair Tommy Koh of Singapore, to negotiate the most contentious and prominent outstanding issues.[59]

Interspersed throughout the UNCED prepcoms and INC sessions were a variety of official and quasi-official meetings addressing questions related to UNCED. For example, ministers from developing countries met in April 1992 at Kuala Lumpur, Malaysia, to agree on a common set of positions for the Rio conference. The Kuala Lumpur Declaration states that development is a fundamental right of all peoples and countries and that an environmentally sound planet ought to correspond to a "socially and economically just world." According to the declaration, sustainable development should be achieved in conjunction with a "supportive international economic environment, including transfer of new and additional financial resources to developing countries, through distinct and specific mechanisms which are transparent, accountable and with equal representation in decision-making, and modalities for transfer of environmentally sound technology. . . ."[60] The declaration calls for internationally supported measures to enhance and stabilize commodity prices (para. 5), the eradication of poverty (para. 6), a transition to more environmentally sound consumption patterns in the North and an interpretation of sustainable development implying the right to development (para. 7), and a separate fund for the implementation of Agenda 21 that would provide new funds in addition to and separate from existing commitments for ODA from developed countries and that would be based on criteria of transparency, democratic and equitable voting, no conditionality, and emphasis on the funding of priorities and needs of developing countries (para 12).

With the blessing of UNCED Secretary General Maurice Strong, a group of "Eminent Persons" met in Tokyo just before the Earth Summit. The Tokyo Declaration, though in no way binding on UNCED delegates, considered questions of international equity and was helpful in diffusing North-South tensions over financing that had resulted from the fourth prepcom.[61] The declaration called for the eradication of world poverty and for the developed countries to reduce their burden on the global environment. It summarized the demands of the developing countries: increased access to markets of the North; increased private investment and technology transfer from the North; solutions to the debt problems of the South, a prerequisite to sustainable development; and "very substantial external support" to complement these efforts. The declaration also called for a restructuring of the Global Environment Facility to make it "fully representative in its decision-making process and transparent in its operations."[62]

Conclusion

International equity has grown in prominence in international environmental discourse and agreements. To be sure, progress on fulfilling and implementing the equity provisions of the UNCED agreements and conventions has been and will continue to be, at least in the near-term, rather limited. Nevertheless, as Henry Shue points out, it appears that international equity is firmly established on the international environmental agenda:

Although it employs the briefest of shorthand, it is worth reminding ourselves that Stockholm 1972 was about Environment, period, and Rio 1992 was about Environment and Development. Between 1972 and 1992 there was a largely unsuccessful attempt to demand a New International Economic Order and a relatively successful attempt to entrench the right to development. The right to development helped to ground the notion of sustainable development, which defined the limits of Rio 1992. Rio had to be, given now dominant norms, about protecting the environment while enabling the poorest to develop. No delegation proposed at Rio that either goal, economic development or environmental protection, should be abandoned; all proposals were about how to balance the two.[63]

Environmental changes (real and anticipated) have acted as stimuli for considerations of equity in international relations.

Increasingly, both North and South agree that the North is inordinately responsible for historic global pollution, that development is understandably more important to the South in the short-term than is environmental protection, that the developing countries will require new funds and technologies provided on concessional terms by the North to assist them in developing in an environmentally benign manner, that the South ought to have control over its resources and ought to have a say in international environmental funding institutions, and that environment and development are inextricably linked issues, with "sustainable development" being the operative term for both environmental and developmental programs. The next chapter distills from the evolutionary process described in this and the previous chapter these emerging themes of international environmental equity. It also introduces and summarizes the U.S. government's posture toward these emerging themes.

Notes

1   Oran R. Young, *International Governance* (Ithaca: Cornell University Press, 1994), pp. 48-50.

2   Evan Luard, *The United Nations* (New York: St. Martin's Press, 1994), p. 160.

3   In the 1960s the North, dominated by the United States, was trying to define North-South relations in terms of the cold war, whereas the South was already attempting to emphasize that North-South relations should revolve around economic issues. See, for example, Marian H. Marchand, "The Political Economy of North-South Relations," in Richard Stubbs and Geoffrey R.D. Underhill, eds., *Political Economy and the Changing Global Order* (New York: St. Martin's Press, 1994), pp. 289-301.

4   See United Nations, General Assembly Resolution 3201, "Declaration on the Establishment of a New International Economic Order," May 1974, and General Assembly Resolution 3281, "Charter of Economic Rights and Duties of States," December 1974.

5   Some very limited successes in calls for technology transfer were achieved during UNCTAD negotiations on an international code of conduct for technology transfer.

6   The United Nations set a target of 0.7 percent of developed country GNP for development assistance. Although several countries did meet or exceed this target, by the mid-1980s the developed countries gave on average only half this amount. For discussion, see Shridath Ramphal, *Our Country the Planet* (Washington: Island Press, 1992), p. 188.

7   The Lomé Convention was signed in 1975 by the European Community and 46 developing countries. The EC granted non-reciprocal duty-free access to all exports of manufactured goods and tropical agricultural produce from those developing countries, and a stabilization scheme ensured that the developing countries' earnings from those exports would remain stable. The agreement also provided for technology transfer and financial and technical assistance. By 1990, 69 developing countries were party to the Convention.

8   Seyom Brown, *International Relations in a Changing Global System* (Boulder: Westview Press, 1996), p. 82.

9   In stark contrast to the objectives of the NIEO, as a consequence of falling commodity prices, reduced development assistance, and ineffective development strategies (frequently mandated by the World Bank and other lenders), by the early 1980s the developing countries as a group had fallen so far into debt that there was a net outflow of capital resources from South to North on the order of $50 billion. Philip Shabecoff, *A New Name for Peace: International Environmentalism, Sustainable Development, and Democracy* (Hanover, NH: University Press of New England, 1996), p. 53; Ramphal, p. 184.

10   The Soviet bloc boycotted the conference because the German Democratic Republic was not allowed to participate.

11   Lynton K. Caldwell, *International Environmental Policy* (Durham: Duke University Press, 1990), p. 56.

12   Gareth Porter and Janet Welsh Brown, *Global Environmental Politics* (Boulder: Westview Press, 1991), p. 130.

13   See Marian A.L. Miller, *The Third World in Global Environmental Politics* (Boulder: Lynne Rienner, 1995), p. 8.

14   Declaration of the United Nations Conference on the Human Environment, UN Doc. A/CONF.48/14 (1972).

15   The United States was opposed to the notion of "additionality," by which developing countries would receive new funds to help them protect the environment, which the government viewed as a requirement for an increase in its foreign aid budget.

16   United Nations, *Law of the Sea: Official Text of the United Nations Convention on the Sea* (New York: United Nations, 1983).

17   The United States was especially concerned that the mobility of its nuclear ballistic missile submarines might be constrained by the extension of coastal states' territorial jurisdictions. This was one of the primary motives for convening the Third UN Conference on the Law of the Sea in 1973. David L. Larson, *Security Issues and the Law of the Sea* (New York: University Press of America, 1994), pp. 126-28; Richard G. Darman, "The Law of the Sea: Rethinking U.S. Interests," *Foreign Affairs* 56, 2 (January 1978), pp. 375-95; Bruce A. Harlow, "UNCLOS III and Conflict Management in Straits," *Ocean Development and International Law* 15, 2 (1985), pp. 197-208.

18 Larson, *Security Issues and the Law of the Sea*, p. 128; see also Henry A. Kissinger, "International Law, World Order and Human Progress," *Department of State Press Release* 4808 (11 August 1975); and Kissinger, *The Law of the Sea* (New York: Foreign Policy Association, 1976), pp. 16-20.

19 Stephen D. Krasner, *Structural Conflict: The Third World Against Global Liberalism* (Berkeley: University of California Press, 1985), p. 230.

20 Exclusive economic zones range from 200 to up to 350 nautical miles, thus excluding vast portions of the world's oceans—including most areas with exploitable resources—from the areas designated as the "common heritage of mankind."

21 United Nations Convention on the Law of the Sea, UN Doc. A/CONF.62/122 (1982), Annex III, Art. 5.

22 David L. Larson, "The Reagan Rejection of the U.N. Convention," *Ocean Development and International Law* 14, 4 (1985), pp. 337-61.

23 William J. Clinton, "The Law of the Sea Convention: Letter to the Senate," *U.S. Department of State Dispatch* 5, 42 (17 October 1994). See the Law of the Sea, "Part XI: Agreement on Deep Sea-Bed Mining." See also Steven Greenhouse, "US, Having Won Changes, Is Set to Sign the Law of the Sea," *New York Times*, 1 July 1994, p. A1.

24 Charlotte de Fontaubert, staff member of the U.S. Senate Committee on Foreign Relations, letter to the author, 29 November 1995.

25 For analyses of the U.S. role in negotiating the ozone treaties, see Richard Elliott Benedick, *Ozone Diplomacy* (Cambridge, MA: Harvard University Press, 1998).

26 Other countries joining this argument included Argentina, Brazil, Egypt, Kenya, Morocco, and Venezuela. Miller, *The Third World*, p. 73.

27 Ibid., p. 71.

28 Alan D. Hecht and Dennis Tirpak, "Framework Agreement on Climate Change: A Scientific and Policy History," *Climatic Change* 29 (1995 [draft of 18 October 1994]), p. 389. The U.S. statement on the Second Meeting of the Conference of the Parties to the Montreal Protocol (London, 6 June 1990) said that "any financial mechanism set out here does not prejudice any future negotiations the Parties may develop with respect to other environmental issues." Ibid., note 45, p. 402.

29 London Amendments to the Montreal Protocol, preamble, para. 7.

30 Art. 5, para. 1 and 5.

31 Art. 10, para. 1. The notion of incremental costs is described in Amanda Wolf and David Reed, *Incremental Cost Analysis in Addressing Global Environmental Problems* (Washington: World Wide Fund for Nature-International, June 1994).

32 Art. 10, para. 3(a).

33 Art. 10A, para. b.

34 The progress of negotiations regarding the Interim Multilateral Fund (including efforts by the South to end its interim designation and efforts by the North to transfer its duties to the GEF) are described in Thomas Gehring and Sebastian Oberthur, "Montreal Protocol: The Copenhagen Meeting," *Environmental Policy and Law*, 23, 1 (1993), pp. 6-12, especially pp. 9-11. The decision of the Copenhagen meeting establishing the full Multilateral Fund is reprinted in *Environmental Policy and Law*, 23, 1 (1993), pp. 51-53.

35 Secretariat for the Multilateral Fund for the Implementation of the Montreal Protocol, "General Information," <http://www.unmfs.org/general.htm>. See also, United Nations Environment Program, "Report of the Thirty-First Meeting of Executive

Committee of the Multilateral Fund for the Implementation of the Montreal Protocol," UNEP/OzL.Pro/ExCom/31/61, 7 July 2000.

36   After the Copenhagen amendments to the Montreal Protocol, developed countries were required to phase out production of CFCs by the start of 1996 and further restrict other ozone-destroying chemicals. UNEP, "Copenhagen Amendment on Ozone Layer to Enter Into Force," Press Release, Nairobi, 22 March 1994.

37   See Jay Shulkin and Paul Kleindorfer, "Equity Decisions: Economic Development and Environmental Prudence," *Human Rights Quarterly* 17 (1995), pp. 382-97, for an examination of the consequences of the Multilateral Fund for equity.

38   Quoted in Philip Shabecoff, "Suddenly, the World Itself Is a World Issue," *New York Times*, 25 December 1988.

39   Communiqué of the meeting of the Group of Seven industrialized countries, Paris, 16 July 1989, reprinted in *New York Times*, 17 July 1989.

40   Marian A.L. Miller, "The Third World Agenda in Environmental Politics," in *The Changing Political Economy of the Third World*, ed. Manochehr Dorraj (Boulder: Lynne Rienner Publishers, 1995), p. 249.

41   "Second Session of Preparatory Committee," *Environmental Policy and Law* 21, 2 (1991), p. 45.

42   Miller, *The Third World*, p. 9.

43   World Commission on Environment and Development (Brundtland Commission), *Our Common Future* (Oxford: Oxford University Press, 1987).

44   The notion of sustainable development is subject to widespread dispute. See, for example, Sharachandra M. Lee, "Sustainable Development: A Critical Review," *World Development* 19, 6 (June 1991), pp. 607-21.

45   Brundtland, p. 43.

46   Ibid, p. 3.

47   United Nations, General Assembly Resolution No. 44/228, 22 December 1989.

48   Nicholas A. Robinson, ed., *Agenda 21 and the UNCED Proceedings* (New York: Oceana Publications, 1992), pp. xvii and lxxxv.

49   Para. 5.

50   Para. 9.

51   Para. 15(m).

52   Preamble, final para.

53   Para. 15(l).

54   The prepcom schedule was as follows: prepcom I organizational meeting, New York, 5-16 March 1990; prepcom I substantive meeting, Nairobi, 6-31 August 1990; prepcom II, Geneva, 18 March-5 April 1991; prepcom III, Geneva, 12 August-4 September 1991; prepcom IV, New York, 3 March-3 April 1992. Because negotiations, especially those regarding institutions, financing and technology transfer, continued during the Earth Summit, the Rio conference is often referred to as "prepcom V." Climate change negotiations began in 1988 with the Intergovernmental Panel on Climate Change (IPCC) meetings and continued with the INC for a Framework Convention on Climate Change (INC/FCCC). Biodiversity negotiations also began in 1988. Selected United Nations documents produced by the pre-Rio negotiations and the Earth Summit are collected in Robinson, *Agenda 21 and the UNCED Proceedings* and Stanley P. Johnson, *The Earth Summit* (London: Graham and Trotman, 1993).

55   Editorial, *Environmental Policy and Law* 22, 4 (August 1992), p. 199. It is worth noting that participation of the poorest of the developing countries was made possible

by funds from the United Nations (donated by developed countries) to help them pay the costs of their delegates' attendance at UNCED-related conferences.

56 "Second Session of Preparatory Committee," p. 45.

57 Miller, "The Third World Agenda," p. 249.

58 Shabecoff, *A New Name for Peace*, p. 138.

59 The entire UNCED negotiating process is summarized in "A Summary of the Proceedings of the United Nations Conference on Environment and Development 3-14 June 1992," *Earth Summit Bulletin* 2, 13 (June 1992), <http://www.iisd.ca/linkages/vol02/0213001e.html>.

60 "Kuala Lumpur Declaration on Environment and Development," Second Ministerial Conference of Developing Countries on Environment and Development, 26-29 April 1992, para. 4.

61 Johnson, p. 31.

62 "Tokyo Declaration on Financing Global Environment and Development," A/CONF.151/7, 4 June 1992 (reprinted in *Environmental Policy and Law* 22, 4 [1992], pp. 265-267). Other meetings that influenced the UNCED outcome included: Aspen Institute Working Group on International Environment and Development Policy, July 1991; Meeting of OECD Ministers of Environment and Development, Paris, December 1991; High-Level Meeting of the members of the DAC [Development Assistance Committee] (joined by the World Bank, IMF, and UNDP), December 1991; IMF and World Bank Development Committee meeting, April 1992; and the GEF Participants' Meeting, May 1992. See Pamela Chasek, "The Negotiating System of Environment and Development" in Bertram I. Spector, Gunnar Sjostedt and I. William Zartman, eds., *Negotiating International Regimes: Lessons Learned from UNCED* (London: Graham and Trotman, 1994), pp. 21-41.

63 Henry Shue, "Ethics, the Environment and the Changing International Order," *International Affairs* 71, 3 (1995), p. 455.

PART II

INTERNATIONAL ENVIRONMENTAL EQUITY AND U.S. FOREIGN POLICY

5 America's Response to Evolving Themes of International Environmental Equity

The evolution of equity in international environmental politics over the last three decades features several themes: (1) the North's inordinate responsibility for historic pollution; (2) the priority of development in the South; (3) increased funding and technology transfer from North to South; (4) greater property and voting rights for Southern countries; and (5) the overriding concept of sustainable development. A limited international consensus has emerged around these themes, which the United States started to join, at first very reluctantly under President Bush, but more positively and sometimes proactively under President Clinton. This chapter will summarize the emerging consensus and the limited acceptance by the Bush and Clinton administrations of the major emerging themes of international equity as objectives of global environmental policy. Subsequent chapters endeavor to explain the U.S. government's posture under Bush and Clinton toward international environmental equity.

Beginning with the Presidency of George Bush, there was a noticeable and historically significant shift in U.S. government policy on issues of environment and international equity. The U.S. government started to move, albeit reluctantly, into line with the emerging international consensus on international environmental equity. Yet, the early Bush administration adamantly opposed almost all of the equity concepts that have been described in this book. The Bush administration was opposed to "new and additional" funding from North to South for sustainable development, concessional technology transfer (and what was perceived as its most potent manifestation, the biodiversity treaty), the sharing of decision making authority in international environmental funding arrangements like the GEF, and "common but differentiated" responsibilities, especially when that concept was interpreted to mean targets and timetables for reductions of U.S. greenhouse gas emissions or related downward adjustment to consumption patterns in the United States.

However, while it never enthusiastically accepted the international equity provisions of the UNCED agreements and conventions, the Bush administration's policy just before, during and after the Earth Summit was clearly less obstructionist, often benign (at least compared to its pre-Rio diplomacy), and on a few issues even supportive.

The Clinton administration, while clearly not as forthcoming as the developing countries and most development and environmental nongovernmental organizations would have liked, was substantially more willing to accept many of these considerations of international equity as valid objectives of effective global environmental policy. In contrast to President Bush, as Democratic presidential and vice-presidential candidates, Governor Bill Clinton and Senator Al Gore campaigned on environmental issues and expressed support for the Earth Summit and many of the equity provisions contained in its products. To be sure, the early Clinton administration was only lukewarm in its actual support of these measures. However, reflecting a shift not unlike that during the Bush administration—but starting from a more sympathetic position—the Clinton administration shifted toward a partial embrace of international environmental equity. In short, the Clinton administration appears to have been more willing to support global efforts to fairly and equitably share the burdens, benefits and decision making authority associated with international environmental issues. As a consequence, during the Clinton administration international equity considerations subtly permeated the entire spectrum of U.S. global environmental policies.

Responsibility of the North

Most governments participating in UNCED embraced the principle that it is the special responsibility of the developed countries to undertake and support efforts to restore and protect the global environment. This allocation of responsibility is based on the developed countries' contribution to global pollution, primarily through industrialized economic development and contemporary high standards of living—in short, the North's high level of consumption. For example, energy use is tied to carbon dioxide emissions, which contribute to climate change. Energy use per capita in the industrialized countries is over thirty times that in almost all developing countries—and one hundred times that in the least developed countries.[1] The recognition of the greater responsibility of the developed countries also reflects the much greater capability of the North to pay for environmental protection efforts.[2] While all countries share the burdens of reconciling environment and development, it is now widely understood that those with the greatest abilities to act must do so first, and help others to act as well.[3] The emerging consensus, reflected at the Earth Summit, is that those countries that have caused suffering should first refrain from causing more harm, and those that are able to do so should take measures to reimburse or otherwise assist those countries that have suffered. Those who have suffered should not have to pay to remedy the consequences of past harm, or to prevent continuing harm, caused by others.[4] This notion is codified in the near-universal agreement among countries on the theme of "common but differentiated responsibility," as stated in Principle 7 of the Rio Declaration.[5]

The Bush administration was generally opposed to any language in UNCED agreements or treaties that might legally establish U.S. responsibility for past pollution or its contemporary global consequences.[6] While the Bush administration did sign on to such language in the Rio Declaration and agreements, it did not embrace the notion that the developed countries—and especially the United States—should bear the brunt of efforts to protect the global environment in the near- and medium-term.[7] Similarly, in the case of the climate change deliberations, during the prepcoms the United States was the only industrialized country refusing to negotiate targets and timetables for controlling greenhouse gas emissions.[8] President Bush's participation in the Rio meeting was made contingent on the FCCC not requiring specific targets and timetables for reduction in emissions of greenhouse gases. According to Philip Shabecoff, "Adamant opposition to the inclusion of binding targets and timetables for limiting the emission of carbon dioxide was the *sine qua non* of the U.S. negotiating posture" at the climate negotiations before the Earth Summit.[9] After the Earth Summit—in the final few months of his effort to be reelected—Bush declared strong U.S. support for the treaty as agreed and pushed it through the U.S. Senate, making the United States the first developed country to ratify the FCCC. However, while he called on countries to formulate their national plans for reducing greenhouse gas emissions and his government developed a voluntary plan to do so, he remained opposed to *binding* targets.

As late as the fourth UNCED prepcom held only weeks before the Earth Summit, the U.S. delegation—alone among the developed countries—insisted on the removal of references in UNCED documents referring to high Northern consumption patterns that have a disproportionate impact on the global environment. It opposed language in the Rio Declaration stating that countries have "common but differentiated responsibilities" and that the developed countries have special responsibilities because of their past consumption patterns and their control of technological and financial resources. To emphasize his opposition to reducing U.S. consumption, shortly before the Earth Summit President Bush said that "the American way of life is not negotiable," making it nearly impossible for U.S. UNCED delegates to make concessions with regard to consumption patterns.[10]

In contrast, the Clinton administration was more forthcoming with regard to Northern responsibility for global pollution and environmental change, at least in its rhetoric. As pointed out in the first chapter, in June 1993, Vice President Al Gore told the United Nations Commission on Sustainable Development that the United States and other developed countries,

> have a disproportionate impact on the global environment. We have less than a quarter of the world's population, but we use three-quarters of the world's raw materials and create three-quarters of all solid waste. One way to put it is this: A child born in the United States will have 30 times more impact on the earth's environment during his or her lifetime than a child born in India. The affluent of the world have a responsibility to deal with their disproportionate impact.[11]

It is on the issues of "common but differentiated responsibilities" and consumption patterns that there was the biggest shift between the Bush and Clinton administrations, according a U.S. diplomat involved in UNCED and its follow-up. Especially with Vice President Gore's influence, the U.S. government accepted the idea that consumption patterns in the North were excessive and ought to be changed, whereas the Bush administration position was to resist such assertions and to push for language in the UNCED agreements that U.S. negotiators thought protected the United States from really having to do anything.[12]

On climate change, the Clinton administration promptly committed the United States to reducing its greenhouse gas emissions to 1990 levels by the year 2000. Policies to achieve this objective were largely voluntary in nature, and they ultimately failed to achieve this goal—much as most other developed countries fell short of the goal. However, in what was uniformly interpreted as a U-turn in U.S. policy, in July 1996 Under Secretary of State Timothy Wirth, representing the U.S. government at the plenary meeting of the second Conference of the Parties to the Climate Change Convention held in Geneva, said that the Clinton administration would support a legally binding instrument or protocol to the convention with specific targets and timetables for reductions of greenhouse gas emissions. The Geneva Declaration, which was agreed at that conference and reminded parties to the FCCC of the convention's principles of equity, common but differentiated responsibility, respective capabilities of Parties, the precautionary principle, and development priorities (not to mention recognition that there was discernible human influence on climate), was "wholeheartedly" endorsed by the United States.[13] At the sixth conference of the parties in November 2000, the United States tried to maximize its freedom in using "flexible mechanisms," like carbon trading and agricultural sinks, to ease its implementation of the Kyoto Protocol. While it arguably spoiled that conference, it never said it was *not responsible* for its inordinate share of greenhouse gas emissions. The environmentalists had already won that argument, and there was no going back. Talk remained focused what to do about it.

Priority of Development

While the objective of the North in international environmental deliberations was, until recently, to address global environmental problems, the goal of the South was to raise living standards. Developing countries wanted economic growth, preferably in conjunction with clean air and water, even at the expense of damage to the larger global environment, at least in the short term. For many countries, local pollution was worth tolerating if it meant economic growth, and for some it was even viewed as a sign of economic success. The emerging international consensus is that the developing countries—especially the least developed countries—have the right to raise the living standards of their populations, and that they ought not be unnecessarily distracted from this objective by requirements to undertake costly programs to protect the global environment, especially when the North bears the greatest responsibility for global pollution to date. The Earth Summit shows that in the effort to balance economic development and environmental protection, the developing countries will often give priority to economic development. The Summit affirmed that both North and South prefer development that does not degrade the local and global environments, but that realizing this objective in the South will require considerable efforts on the part of the North.

The U.S. government has supported economic development— usually described as economic growth—in developing countries. Indeed, it would have been surprising if the United States took a contrary position in light of the advantages that could accrue for U.S. exporters as a result of economic development in the South. The Bush administration recognized that poverty eradication and economic development were the primary priorities of the developing countries in the UNCED negotiations. The chief U.S. UNCED diplomat told Congress three months before the Earth Summit that "the clear message that has come out of all our deliberations on UNCED is that poverty is the number one issue facing the developing world. And unless we, in the UNCED process within Agenda 21 address this issue, find means to alleviate poverty, we are not going to accomplish anything."[14] The Bush administration's support for development fell short of acknowledging a *right* to it, but the U.S. delegation to the Earth Summit chose not to force a renegotiation of the Rio Declaration's language on the "right to development" (Principle 3). Instead, the United States issued an "Interpretive Statements for the Record" in which it stated its opposition to such a right.

The Clinton administration recognized the importance of sustainable development in the South, both for the good of those countries themselves and for U.S. economic and security interests. It focused substantial diplomatic resources on poverty eradication programs, especially in the least developed countries, and it requested financial resources from Congress for that purpose. The Agency for International Development, formerly focusing much of its resources on development programs with cold war motivations, restructured its aid programs to focus on sustainable development in the South. However, while the U.S. government under Clinton supported many of the sustainable development goals of the developing countries and recognized that economic development is their priority, it did so while emphasizing the advantages of free-market measures and the work of nongovernmental organizations and local communities, rather than reliance on substantial new foreign aid for programs administered by national governments.

New and Additional Funds

There was and continues to be extensive disagreement between the affluent and the poor countries over how to finance the environmental agreements emanating from the Earth Summit. Most developed countries are not very interested in providing *new* funds for developing countries, as reflected in the fall in official development assistance during the 1990s. They would prefer to restructure existing aid programs to better serve environmental objectives. For their part, the developing countries have always insisted that they cannot and ought not be required to fund programs to protect the global environment, and even their own local environments in many cases, because they simply do not have the funds to do so and because they are not responsible for global environmental problems. To bridge these differences, countries have agreed, at least in principle, that new and additional funds should be allocated to finance the "incremental" costs of international environmental agreements like those made in the context of UNCED.[15]

The Bush administration consistently refused to acknowledge any responsibility for providing new and additional funds to developing countries for sustainable development—or any other purpose. The U.S. government was initially adamantly opposed to an international funding institution to support sustainable development in the South. Thus, it was initially opposed to the GEF, but preferred it to a "green fund" that the developing countries were demanding. At the third UNCED prepcom the one central theme that dominated and influenced all aspects of U.S. participation in the deliberations was the U.S. opposition to new and additional funding for environment and development.[16] Similarly, at the fourth prepcom the United States refused all concessions to the developing countries' demands for new and additional financial resources for sustainable development, although it did compromise by agreeing to accept the "new and additional" language while interpreting it as meaning that new funds for environmental protection would come from existing assistance. Despite the Bush administration's general opposition to such funds, and its continued skepticism of the need and efficacy of such funding, in 1990 Secretary of State James Baker said: "We fully recognize that developing countries may need some additional aid in order to meet the incremental costs associated with fulfilling their international environmental obligations."[17] The head of the U.S. UNCED delegation, EPA administrator William K. Reilly, declared at the Earth Summit that "Whether we are able to realize the goals we set for ourselves at this conference will also depend on technology cooperation and financial assistance. . . . we recognize the need for outside resources to assist developing nations."[18] The Bush administration hardly embraced new funding, but it did agree that new funds would have to come from somewhere.

The Clinton administration supported, at least rhetorically, the UNCED provisions calling for new and additional funding to help developing countries undertake sustainable development. It increased funding to the GEF and was more forthcoming generally in trying to provide financial support for international environmental and developmental funding mechanisms, especially when compared to the real reluctance of the Bush administration to do so.[19] However, the administration was unable to garner substantial new foreign aid funds from Congress. Instead, it directed scarce development assistant resources to sustainable development programs, with a new focus on poverty eradication in the least developed countries along with more efficient use of resources.[20] Importantly, the United States has never recognized the UN goal of providing 0.7 percent of GNP for development assistance. It is unlikely to do so in the foreseeable future, primarily because Congress is unwilling to relinquish its periodic authority to set levels of foreign aid.[21] However, suggesting that the executive and legislative branches can agree on new aid, the Clinton administration and the Congress agreed in 2000 to join other countries in renouncing the massive debt of many of the world's poorest countries.

Technology Transfer

At almost every international conference to address global environmental problems the developing countries have demanded that appropriate technologies be transferred to them on noncommercial or concessional terms.[22] They have faced intense opposition from several industrialized countries, especially the United States. But the United States has become increasingly isolated in its opposition to technology transfer. The amendments to the Montreal Protocol contain important provisions to help the developing countries acquire chemicals that will replace ozone-damaging CFCs. And the UNCED declarations and conventions contain extensive provisions for technology transfers to help developing countries take steps to protect their forests and biological resources and to limit climate change. If the South is going to develop with or without help from the North, then it behooves the industrialized countries to provide them with technologies that permit development with less harm to the environment. Such was the consensus at Rio, although how to operationalize this consensus remains extremely problematic.

While the U.S. government has supported the general notion of technology transfer (what the Bush administration preferred calling "technology cooperation"), it has been adamantly opposed—especially during the Bush administration—to international agreements that *require* transfer of technology on concessional terms. The Bush administration's William Reilly said that the "United States strongly supports technology cooperation with developing countries to help them find sustainable paths to economic development,"[23] and President Bush acknowledged that developing countries "will need new technologies if they are to enjoy 'green growth.'"[24] However, the Bush administration was quite consistent in opposing policies that require U.S. businesses to transfer technology at less than market rates or that appear to threaten intellectual property rights of U.S. corporations. Largely in response to U.S. objections, the Agenda 21 chapter dealing with technology transfer contained hedged language that called for "promoting" rather than guaranteeing developing country access to environmentally sound technology, and contained references to "measures to prevent the abuse of intellectual property rights" of developed country corporations.[25] Such concerns were the administration's primary justification for refusing to sign the Convention on Biological Diversity.

The Clinton administration was somewhat more supportive of efforts to get new environmental technologies to developing countries. Toward that end, it signed the Biodiversity Convention in June 1993. In so doing, it attached an interpretive statement asserting that the convention's provisions dealing with intellectual property rights (as well as the financial mechanism) were "compatible" with U.S. interests.[26] However, the United States has not ratified the treaty, nor is it likely to do so soon given opposition in the Senate.[27]

Property Rights

The South has garnered additional international rights over the use and fruits of natural resources. In the case of the deep seabed, traditionally owned by no person or state, the developing countries now have rights to a portion of profits that might accrue from its exploitation. Regarding their own forest and especially genetic resources, they have established sovereignty over use and access, as well as the right to benefit from biotechnologies that derive from those resources.[28]

The United States has been slow to recognize collective property rights to commons areas. The U.S. government was central to the Law of the Sea compromise that traded rights of passage for declaration of the deep seabed as the "common heritage of mankind," but the Reagan and Bush administrations were opposed to that declaration, preferring that access to the deep seabed go to private corporations able to exploit resources there. The Clinton administration ended long-standing U.S. opposition to the Law of the Sea after renegotiation of the deep seabed provisions to reduce the costs of corporate access to that area. Nevertheless, the deep seabed remains the "common heritage of mankind," with all the implications that holds for future negotiations regarding global commons.

With regard to national property rights, the Bush administration recognized that countries ought to benefit from their own resources. William Reilly said at Rio that "we [the United States] do think that countries need to be rewarded for the resources that they make available to help improve the quality of life here in the world."[29] The Clinton administration recognized all countries' rights to control their own natural and genetic resources and to benefit from them, provided that mutually agreed and established rights of U.S. corporations did not suffer. It thus went the next step beyond the Bush administration by signing the Biodiversity Treaty.

Voting Rights

In international environmental funding institutions like the ozone Multilateral Fund and the Global Environment Facility, developing countries now have a greater portion of the decision making power than they would have if such rights followed the custom of allocating votes according to contributions—and thus to the wealthiest developed countries. The conferences of the parties to the climate change and biodiversity conventions also have much or most of the power to determine funding priorities. Thus, the developing countries are increasingly able to influence decisions regarding which among them will benefit from funds donated by the developed countries.

The Bush administration successfully pushed for the Global Environment Facility as the funding mechanism for global environmental programs (because it was opposed to the developing countries' alternative, a "green fund" under United Nations control). Although it initially opposed such measures, the United States, alone among the developed countries, was the first developed country to propose that the GEF should be "democratized"[30] in order to make the facility more palatable to the developing countries.[31] The Clinton administration implemented this policy, supporting the post-Rio "democratic" restructuring of the GEF that has given more authority to developing countries.

Sustainable Development

The North wants to protect the global environment because it is in its interest to do so. The South, while also having great interest in protecting the global environment, has chosen to prioritize economic development, in large part due to the poor standards of living under which many of its people live. The notion of sustainable development, popularized by the Brundtland Commission, is one theme that permits a convergence of these two perspectives. An explicit component of the Brundtland Commission's definition of sustainable development is the need to reduce poverty. The emerging international consensus holds that environmental destruction can contribute to poverty and is caused by it; wise economic development can lift people out of poverty while simultaneously protecting the environment. According to this view, both the North and the South have an interest in protecting the planet's environment, just as both have an interest in economic development to limit global poverty. The paradigm of the Industrial Revolution that fueled development in the North is no longer compatible with long-term planetary survival, and environmental vulnerability requires a transition to sustainable development. The concept of sustainable development thus has provided a rudimentary "conceptual bridge" for the "critical partnership between North and South [required] for addressing the problems of environment and development."[32] By focusing on promoting sustainable development, as the Earth Summit did, all countries have begun to see that they have common interests, with developed countries beginning to recognize that it is in their interest to promote international equity in order to protect those interests.

The United States spent many months opposing the title of the Earth Summit's Rio Declaration on Environment *and* Development due to the implication that environmental protection and economic development should be linked and co-equal. The United States preferred that the declaration be called the "Earth Charter," a title that suggested environmental protection as the priority of the summit. However, despite U.S. opposition during much of the negotiations, the Rio Declaration reflects the development agenda of the South much more than did the Stockholm Declaration signed twenty years before at the UN Conference on the Human Environment. While the Bush administration hesitated to make the environment/development connection part of U.S. foreign policy, there was clear recognition in the Bush administration that economic development and environmental protection were linked in practice, and that sustainable development was an important objective. Indeed, Secretary of State Baker said that "Sustainable development . . . is a way to fulfill the requirements of the present without compromising the future. When policies of sustainable development are followed, our economic and environmental objectives are both achieved. . . . America's entire approach to bilateral and multilateral assistance is based on the concept of sustainable development."[33] Discussing the objectives of the Earth Summit, President Bush stated: "I think it's critical that we take both those words—environment and development—equally seriously, and we do."[34] Environmental Protection Agency administrator Reilly said that "the United

States embraces enthusiastically the goals of this conference, its priority to sustainable development and to a better relationship between humans and nature. . . ."[35]

Demonstrating the U.S. government's awareness of the environment-development connection, the U.S. delegation to the third UNCED prepcom distributed a statement saying that the United States was "strongly committed . . . to combating poverty and famine and to raising global living standards. We focus our attention and resources on precisely those factors which cause both environmental degradation and underdevelopment—population pressure, economic, political, and social inequities, and declining agricultural productivity. And in all of our assistance activities, we strive, in the words of the Brundtland Commission, to meet the needs of the present without compromising the ability of future generations to meet their own needs."[36] After the Earth Summit, the Bush administration policy was that a guiding principle of UN economic and social activities should be to support the "integration of environment and development in order to achieve sustainable development."[37]

With the advent of the Clinton administration, sustainable development became an increasingly important theme in U.S. foreign policy. A 1993 congressional briefing book prepared by the Congressional Research Service described the new prominence of sustainable development as "an evolution for environmental and development perspectives: the newest form of sustainability not only includes environment in development, but also puts a broader perspective on actions that need to be taken if the environment is to be protected—a perspective that includes basic human needs and economic development as a corollary to better environmental management."[38] Building on the Bush administration policy, the Clinton administration adopted sustainable development as a framework for its foreign policy generally and its foreign development policies in particular. In June 1993 the Clinton administration announced the creation of the President's Council on Sustainable Development, a 25-member council made up of representatives from the administration, industry, and environmental groups. Timothy Wirth stated that "the pursuit of sustainable development—the lofty idea launched at the Earth Summit in Rio—must be the pillar of renewed American foreign policy and redefined national security for the 21st century," and that "in the United States and around the globe, we are coming to understand the close connections between poverty, the environment, the economy, and security. This historic transformation demands that we now liberate ourselves—from outworn policies, from old assumptions, from fixed views that only yesterday seemed to be the dividing and defining lines of our politics."[39]

Secretary of State Christopher, describing the 1995 foreign affairs budget, told the Senate that the Clinton administration was "dedicated to restoring America's leadership role on sustainable development—an approach that recognizes the links between economic, social, and environmental progress. We are putting this global challenge back where it belongs: in the mainstream of American foreign policy and diplomacy."[40] This commitment to sustainable development continued through the second Clinton administration, and was reflected in how the State Department and the Agency for International Development deployed limited funds.[41]

The Case of "Common but Differentiated Responsibility" and U.S. Climate Change Policy

To summarize, the U.S. government has in recent years come to accept that international equity must be given serious consideration in global environmental politics. It has even started, albeit only modestly, to join the emerging consensus on the various themes of international environmental equity outlined above. While I do not wish to overstate the degree to which the United States has accepted and endorsed these themes, it clearly has done so to a degree. This is reflected in the interesting case of "common but differentiated responsibility," a key feature of the emerging norms of international equity in global environmental politics, in the context of the Climate Change Convention.[42]

The overriding goal of the FCCC is "stabilization of greenhouse gas concentrations in the atmosphere at a level that would prevent dangerous anthropogenic interference with the climate system."[43] Industrialized countries agreed in the FCCC to *voluntarily* reduce their emissions of greenhouse gases to 1990 levels by 2000. Very few did so, and the United States probably exceeded this target by at least 13 percent.[44] In an effort to achieve more concrete action on greenhouse gas emissions, parties to the Convention agreed to negotiate, in time for the third conference of the parties in Kyoto at the end of 1997, a protocol laying out binding targets and timetables for reductions of greenhouse gases by developed countries.[45] Among the key principles in the Convention left to be operationalized was that of "common but differentiated responsibility" (CBDR), whereby industrialized developed countries would take the lead in addressing the climate problem, specifically excluding developing countries from binding greenhouse gas emissions reductions.[46] The developed countries are disproportionately responsible for historical greenhouse gas emissions, and they have the greatest capacity to act.[47] Thus the Convention makes few demands on the much less responsible and usually much less capable developing countries. This exclusion of developing countries became one of the most contentious issues before and during the Kyoto conference (and remains so), especially because the United States insisted that developing countries make "meaningful" contributions to future greenhouse gas reduction efforts. These U.S. demands appeared to contradict the CBDR principle.

The first basic principle of the Climate Change Convention, Article 3(1), states that:

> The Parties should protect the climate system for the benefit of present and future generations of humankind, on the basis of equity and in accordance with their common but differentiated responsibilities and respective capabilities. Accordingly, the developed country Parties should take the lead in combating climate change and the adverse effects thereof.[48]

The convention recognizes that all countries are responsible for climate change, and all should endeavor to limit the pollution that causes it. However, following the CBDR principle, the treaty does not require developing countries to reduce their greenhouse gases. It instead requires the developed countries to take the "lead in modifying longer-term trends in anthropogenic emissions [of greenhouse gases] consistent with the objective of the Convention."[49]

The CBDR principle was reaffirmed in 1995 at the first conference of the FCCC parties in Berlin. Countries agreed to the "Berlin Mandate," whereby developed countries pledged to act first to reduce their greenhouse gas emissions before requiring developing countries to do so.[50] The Berlin Mandate declares that the process of implementing the Climate Change Convention shall be guided by, *inter alia*, the CBDR principle, and it quotes Article 3(1) of the Climate Change Convention (see above).[51] It reminds parties that they are required to consider the special needs of the developing countries and:

> The fact that the largest share of historical and current global emissions of greenhouse gases has originated in developed countries, that the per capita emissions in developing countries are still relatively low and that the share of global emissions originating in developing countries will grow to meet their social and development needs.[52]

> The fact that the global nature of climate change calls for the widest possible cooperation by all countries and their participation in an effective and appropriate international response, in accordance with their common

but differentiated responsibilities and respective capabilities and their social and economic conditions.[53]

Thus negotiations for the Kyoto Protocol between 1995 and 1997 were premised on the CBDR principle. The United States, along with other developed country parties to the Convention, accepted this standard because they knew developing countries would not—and in many cases could not—limit their emissions otherwise.[54] But, to many observers, the U.S. policy toward common but differentiated responsibility and climate change appeared to shift in the months before the Kyoto gathering. Indeed, the common interpretation was that the United States had abrogated its responsibility with regard to the CBDR principle.[55]

Throughout the international negotiations on a protocol to the Climate Change Convention, developing countries consistently declared that they would not agree to any limitations (least of all *reductions*) in their greenhouse gas emissions until the developed countries substantially reduce theirs.[56] The developing countries acted on these sentiments in Kyoto, vetoing any language in the Protocol that would call on them to make even voluntary commitments to limit their emissions of greenhouse gases.[57] Accordingly, the Kyoto Protocol, agreed on 10 December 1997, requires *developed* countries to reduce their overall emissions of greenhouse gases by about five percent below 1990 levels by 2012. The United States agreed to reduce its emissions by seven percent, the Europeans by eight, and the Japanese by six. A handful of developed countries, such as Australia, were allowed to increase their emissions.[58]

Conforming to the Climate Change Convention's provisions for common but differentiated responsibility, and specifically reaffirming the Berlin Mandate, the Kyoto Protocol does not require the developing countries to take on new commitments to limit their greenhouse gas emissions. Indeed, the Protocol is devoid of references to commitments of developing countries. Rather, all its provisions apply to the developed "Annex I" countries. In Article 10 the Protocol explicitly reaffirms common but differentiated responsibility when it states that all parties must take into account "their common but differentiated responsibilities and their specific national and regional development priorities, objectives and circumstances, without introducing any new commitments for Parties not included in Annex I [i.e., the developing countries]. . . ."[59]

The U.S. government consistently supported common but differentiated responsibility in the context of climate change, despite contrary interpretations in most press reports.[60] It is true that the U.S. position on common but differentiated responsibility differs from that of other countries, especially if one is concerned with rhetoric from Congress and strict interpretations of proposed treaty wording.[61] This was evident during the months before Kyoto when the Clinton administration called for "new" and "meaningful" commitments from developing countries in the protocol.[62] However, it would be wrong to say that the United States expected developing countries to take on "common" responsibilities instead of "common *but differentiated*" responsibilities. The U.S. government joined the 1995 Berlin Mandate, thereby reaffirming the CBDR provisions of the Climate Change Convention. Even Congress, particularly the Senate, declared its support for common but differentiated responsibility, as indicated by extensive debate on the Senate floor. In substance, the U.S. government accepted and actively promoted the CBDR principle in the climate change negotiations—even if this acceptance was not of the kind that other governments were willing to accept.

The U.S. Senate's Position on Common but Differentiated Responsibility

In July 1997, by a vote of 95-0, the U.S. Senate adopted Senate Resolution 98 (SR-98), the so-called Byrd-Hagel Resolution.[63] That Resolution stated, *inter alia*, that the United States should not be a signatory to any protocol to, or other agreement regarding, the Climate Change Convention that would:

(A) mandate new commitments to limit or reduce greenhouse gas emissions for the Annex I [developed country] Parties, unless the protocol or other agreement also mandate new specific scheduled commitments to limit or reduce greenhouse gas emissions for Developing Country Parties within the same compliance period, or

(B) would result in serious harm to the economy of the United States.[64]

The Byrd-Hagel Resolution reflected concern about the effects of any binding climate treaty on the U.S. economy, and distrust among both Republican and Democratic senators of an agreement that would exclude the developing countries. On first reading, the Resolution sounds ominous to those concerned about common but differentiated responsibility and international fairness, and observers could be forgiven for interpreting it as a "treaty killer." But such an interpretation would miss the Resolution's message. Rather than focus on the rhetoric of the most outspoken anti-Climate Change Convention senators, it is instructive to look at the meaning of the Resolution made explicit by senators in floor debate, as well as the interpretation of it described by the Clinton administration in its public statements. Such an examination shows that the Byrd-Hagel Resolution was based fundamentally on conceptions of common but differentiated responsibility and notions of fair and equitable international burden sharing. While this does not diminish the reality that the United States would not join the robust efforts to reduce greenhouse gases that were required, it does nevertheless show that equity had become an accepted norm, even in the legislative branch of government.

The Senate's concerns about developing country participation in the FCCC process were at least two-fold.[65] First, whether rightly or wrongly, there was concern that developing countries would have an unfair economic advantage because they would not be facing the same restrictions on economic output as the United States (GNP is, for better or worse, still largely proportional to energy use, and thus greenhouse gas—especially carbon dioxide—emissions). There was also the concern that U.S. manufacturing, and hence U.S. jobs, would move abroad to take advantage of relaxed environmental regulations there. Second, the Senate—and the Clinton administration—believed that an effective climate treaty absolutely requires developing country participation.[66] During floor debate on the Resolution, senator after senator cited the future emissions of China as justification for their position. They pointed out that by 2015 China will surpass the United States to become the world's leading producer of greenhouse gases, and they often noted that other large developing countries were rapidly increasing their emissions.[67]

Perhaps the greatest confusion about the Senate's Byrd-Hagel Resolution concerns its provision to "mandate new specific scheduled commitments to limit or reduce greenhouse gas emissions for Developing Country Parties within the same compliance period" as the United States and other developed country parties. During floor debate, several senators went to great lengths to establish the precise interpretation of this provision. Most senators accepted that the U.S. was more responsible for the problem; they accepted that the developing countries, while they should undertake "specific scheduled commitments" in the "same compliance period," should not be required to undertake the same commitments as the developed countries; and they accepted that the poorest and least capable countries should have the least difficult requirements placed upon them (or none at all).[68] Indeed, one could readily interpret the Resolution as accepting that many developing countries would *increase* their greenhouse gas emissions—but that those increases ought to be codified in the agreement. Indeed, this was the U.S. negotiating position at the Kyoto conference.[69]

It is worth citing some senators' comments to demonstrate that the

U.S. Senate accepted the fundamental provisions of the Climate Change Convention and the CBDR principle. Many legislators recognized that the United States was the largest part of the climate change problem, and that the United States should act to reduce its greenhouse gas emissions while assisting developing countries with their actions. For example, in a statement on the Senate floor, Senator Patty Murray said the following:

> Regarding the developed-developing nation debate, I believe it is also clear that we developed nations have historically emitted more greenhouse gases per capita than developing countries. In addition, we are economically more able to absorb whatever increased costs occur based on the need to reduce emissions. Therefore, we should assist our neighbors through technology transfer, economic assistance, and joint ventures in meeting whatever emissions goals are established.[70]

Senator Robert Byrd, primary co-sponsor of SR-98, interpreted the developing country provisions of the Resolution this way: "Now, does this mean that the Senate is insisting on commitments to identical levels of emissions among all the parties? Certainly not. The emissions limitations goals, to be fair, should be based on a country's level of development. The purpose is not to choke off Mexico's development or China's development."[71]

Similarly, during Senate debate, Senator Robert Kerry described interpretations of what he thought the phrase "same compliance period" actually meant:

> [I]t means essentially that we want countries to begin to reduce while we are reducing, we want them to engage in a reasonable schedule while we are engaged in a reasonable schedule, but that if a developing nation needs more time to get a plan in place or needs to have more time to raise the funds and be able to purchase the technology and do the things necessary, that as long as there is a good-faith track on which they are proceeding, that if it took them a number of years . . . to reach a particular goal, that certainly means within the same compliance period . . . . [I]t is reasonable to permit some flexibility in the targets and timing of compliance while at the same time requiring all countries to agree to make a legally binding commitment by a date certain. That is reasonable. But I think most of my colleagues would agree that if some country simply doesn't have the capacity, the plan, the money, or the technology, it may be they have to take a little more time and we should want to be reasonable in helping them do that because the goal here is to get everybody to participate, not to create a divisiveness that winds up with doing nothing.[72]

Senator Max Baucus reiterated the need for the developing countries to have the same compliance period, and added: "But since developing and developed nations are starting from different places, it makes sense to require different targets. Here again, the language crafted by Senator Byrd helps. It does not specify that developed and developing countries meet the same targets and timetables."[73] Senator Kerry echoed Baucus's remarks. He said that the Senate resolution would allow developing countries "appropriate flexibility" in their commitments to abate greenhouse gas emissions.[74] He added that it was the developed countries who were "in a better position to implement emissions-curbing activities and technologies at low cost and impact, and to transfer these abilities and technologies to developing countries and to aid in their economic advancement in a way that tempers emissions growth."[75]

Senator Joseph Lieberman reinforced these interpretations in the following statement:

> New commitments by developing countries regarding their performance under the [FCCC], of course, need to be consistent with their historic responsibility for the problem, as well as their current capabilities. The ground rules for the negotiations—the Berlin mandate—recognize these common, but differentiated responsibilities. It is clear that the Berlin mandate can be carried out in a way that is consistent with Senate Resolution 98. The resolution says that developing countries can start with a commitment that is lower relative to the industrialized countries at first. . . . Senate Resolution 98 says that it is entirely appropriate for industrialized countries to start making quantified emissions reductions first. . . .[76]

To put to rest confusion about the interpretation of the developing country provisions of the Byrd-Hagel Resolution, Senator Jeff Bingaman entered a colloquy with Senator Byrd:

> Bingaman: I was greatly encouraged by the remarks on this issue made by the sponsor of this resolution [who said that] countries at different levels of development should make unique and binding commitments of a pace and kind consistent with their industrialization . . . and consistent with a fair sharing of any burden. . . . Would it be correct to interpret the use of the words "new commitments" in both phrases as suggesting that the United States should not be a signatory to any protocol unless Annex I Parties and Developing Country Parties agree to identical commitments?

> Byrd: That would not be a correct interpretation of the resolution. [I said and] deliberately repeated it for emphasis: "Finally, while countries have

different levels of development, each must make unique and binding commitments of a pace and kind consistent with their industrialization." I believe that the developing world must agree in Kyoto to binding targets and commitments that would begin at the same time as the developed world in as aggressive and effective a schedule as possible, given the gravity of the problem and the need for a fair sharing of the burden. That is what the resolution means. The resolution should not be interpreted as a call for identical commitments between Annex I Parties and Developing Country Parties.[77]

To be sure, it would be much easier for international negotiators and for the Clinton administration if the Senate had never passed the Byrd-Hagel Resolution. Many observers interested in international equity hoped the U.S. would be more forthright in taking on its fair share of the climate change burden. One might even assume that some senators were not interested in international fairness; they might be happy to see developing countries subjected to immediate mandated reductions in their greenhouse gases. Nevertheless, as the senators' statements show, the Resolution was not what it might appear to be; it was not a renunciation of the CBDR principle, but rather an interpretation of it—albeit a much less robust one than the developing countries wanted.

The Clinton Administration's Position on Common but Differentiated Responsibility

Responding to the Byrd-Hagel Resolution, and bearing in mind the developing countries' increasing emissions of greenhouse gases, on October 22, 1997, President Clinton announced his administration's negotiating position for the Kyoto conference.[78] He called on the industrialized countries to commit to a binding and "realistic" target of returning to 1990 emissions of greenhouse gases between 2008 and 2012, reducing emissions below 1990 levels during the subsequent five years.[79] President Clinton declared that "both developed and developing countries must participate in meeting the challenge of climate change."[80] "Developing countries have the opportunity to chart a different energy future consistent with their growth potential and their legitimate economic aspirations," the President added.[81] He said that "key" developing countries (those that are the largest emitters and the wealthiest developing, "non-Annex I," parties, e.g., China, India and Mexico) must take "meaningful" action, but that the "industrialized world must lead."[82] Thus "meaningful participation" of "key" developing countries did not mean "equal participation."

Affirming that the U.S. position was to support, not harm,

economies in the developing world, President Clinton said he would not propose changes to the Climate Change Convention that would adversely affect the growth of developing countries, who feared that emissions reductions might penalize them because their industries are often less pollution conscious.[83] The President said the United States wanted "to help the developing nations grow as much as they would without a treaty, but on a different energy path than the one we charted when we were at the same stage of development."[84] In his 22 October announcement of the U.S. Kyoto negotiating position, Clinton again acknowledged the special U.S. responsibility for the problem: "The United States has less than five percent of the world's population, enjoys 22 percent of the world's wealth, but emits more than 25 percent of the world's greenhouse gases."[85]

In Senate testimony, then Under Secretary of State for Global Affairs Timothy Wirth described the U.S. position on many of the most contentious questions of equity and fairness in the context of climate change.[86] Despite facing hostile questions from some senators, his description of U.S. policy accommodated the CBDR principle, especially with regard to U.S. expectations of the developing countries.[87] Wirth noted that the developing countries would have to be part of the treaty because their greenhouse gas emissions were increasing rapidly.[88] He pointed out at the same time, however, that the developing countries' per capita emissions would continue to remain "far below our own," and he defended the actions that the developing countries had already taken to reduce their greenhouse gas emissions, despite (as he made clear) their relative poverty.[89] Wirth said the United States "must determine what we ask of developing countries with realistic and fair appreciation of how they see the world as well. The level and timing of each country's commitments must be commensurate with its national abilities and level of development. Balance and fairness must be maintained."[90] His remarks, like those of Vice President Al Gore[91] and other administration officials,[92] mirrored the CBDR principle contained in the Climate Change Convention.[93]

The phrasing of the Senate resolution ("limit or reduce") left room for the Clinton administration to agree that developing countries should be able to *increase* their greenhouse gas emissions.[94] That is precisely what the United States diplomats proposed at the Kyoto conference: American diplomats called for *voluntary* commitments by developing countries (specifically excluding the least developed among them) to "abate the increase" in their emissions.[95] The U.S. diplomats said that developing country emissions targets could be *growth* targets, and that commitments should not inhibit economic development in those countries.[96] In other words, the Clinton administration wanted the large developing countries to plan their future emissions and commit themselves to adhering to those plans. The U.S. diplomats only wanted something—virtually anything—in the Protocol's wording that would allow the administration to say to Congress that developing countries were "limiting" their emissions in "meaningful" ways. So the U.S. position was not, as it was billed by almost everyone but the Clinton administration, an abrogation of the CBDR principle. To be sure, the Byrd-Hagel Resolution restrained the Clinton administration's diplomacy, as reflected in U.S. positions taken at the sixth conference of the FCCC parties. However, again, the United States sought to work within the constraints of the CBDR principle. The U.S. government never denied its importance, and indeed at least continued to assert its legitimacy.

Conclusion

By the mid-1990s an international consensus on international environmental equity had begun to emerge. The international community agreed that the North, because of its historically high levels of consumption and global pollution, and because of its wealth, had the primary responsibility for addressing global environmental problems. Alternatively, there was agreement that the South ought not be sidetracked in its efforts to develop economically by expensive measures to protect the environment; economic development would remain more important than environmental protection in the developing world. There was also consensus that the developing countries will require new and additional financial assistance from the developed countries to pay the additional costs the South incurs for environmental protection programs and specifically for programs to meet the provisions of international environmental agreements (e.g., Montreal Protocol as amended, FCCC, Biodiversity Convention). Similarly, there was acknowledgment that the South will require concessional transfers of environmental technologies to help it undertake sustainable development. Consensus also emerged with regard to property rights. Certain common areas (such as the deep seabed) are appropriately the common heritage of humankind, and their exploitation ought to benefit, at least in part, the poorer countries that do not have the financial or technological capacity to exploit those common areas. Alternatively, the resources that are found within a country's borders, even if that country by itself cannot exploit those resources, are its sole property; thus the developed countries (or corporations from them) must get permission to use those resources and should share the fruits of developing those resources with the country of origin. Countries also agreed that the traditional weighted voting of international financial institutions ought to be adjusted in the environmental field to give developing countries the right to participate in the decision making of those institutions. And, most important, consensus has emerged in both North and South that environmentally sustainable development, which ties together environmental protection, economic development and poverty eradication, should be the guiding principle for environmental protection and economic development programs within national jurisdictions, as well as development assistance programs between North and South.

For most of the UNCED preparatory period—from 1989 to early 1992—the Bush administration was not greatly interested in the UNCED deliberations. One critic told Congress that "it appears that USA preparations for UNCED are at best those minimally necessary to keep our seat at the table."[97] According to an observer of the entire UNCED process, "For most of the presummit process, the U.S. position was adamant: No concessions on new and additional resources, no transfer of technology on concessional terms, no concessions on consumption by the rich, no new economic order."[98] While the United States was opposed to many of the equity considerations deliberated in the context of UNCED, its opposition softened shortly before, during, and especially after the Earth Summit. Consequently, other actors more interested in promoting international environmental equity were able to push through the equity provisions of the agreements and conventions described in previous chapters.

In contrast, the Clinton administration endeavored to accept and promote international equity as an objective of global environmental policy. It acknowledged that the United States is among those countries that were inordinately responsible for global environmental pollution and that the developing countries would be unable to deal with the consequences of that pollution without outside assistance. It acknowledged that consumption patterns in the United States and other developed countries had an inordinate impact on the global environment and would require adjustment downward. It acknowledged that the United States would have to take on the burden, before developing countries do, to reduce its emission of greenhouse gases, and it stated its support for a legally binding agreement toward that end. The Clinton administration made sustainable development a central principle of U.S. foreign policy, integrating the concept throughout the entire foreign policy bureaucracy and, where applicable, the defense establishment. Moreover, during the tenure of Clinton even the legislative branch—despite the fear of a climate treaty by many members of Congress—fundamentally endorsed the notion of international environmental equity in the context of the Climate Change Convention.

The U.S. response to the emerging norms of international equity in global environmental politics fell far short of what many reasonable people and many other governments wanted. Nevertheless, it was recognition that equity would have to be an important consideration, and at times the U.S. government even endorsed efforts to make global environmental politics more equitable and fair for the world's poor.

Notes

1. United Nations Development Program (UNDP), *Human Development Report 1994* (Oxford: Oxford University Press, 1994), p. 191.
2. Marian A.L. Miller, "The Third World Agenda in Environmental Politics," in Manochehr Dorraj, ed., *The Changing Political Economy of the Third World*, (Boulder: Lynne Rienner Publishers, 1995), p. 251.
3. Mukal Sanwal, "Sustainable Development, the Rio Declaration and Multilateral Cooperation," *Colorado Journal of International Environmental Law and Policy* 4, 1 (Winter 1993).
4. Cf. Henry Shue, "Equity in an International Agreement on Climate Change," in Richard Samson Odingo et al., eds., *Equity and Social Considerations Related to Climate Change* (Nairobi: ICIPE Science Press, 1995), pp. 385-92.
5. For additional discussion of this theme, see, for example, Daniel Bodansky, "Draft Convention on Climate Change," *Environmental Policy and Law* 22, 1 (1992), p. 12.
6. In contrast, Bush's conservative counterparts in the United Kingdom—specifically Prime Minister Margaret Thatcher—felt that the North had special responsibility to take action to protect the global environment for future generations. Chris Thompkins, Minister's Representative, Environmental Protection International, Department of the Environment, United Kingdom, telephone interview by author, 7 August 1996.
7. The U.S. issued an interpretive statement of Principle 7 of the Rio Declaration in which it rejected any interpretation that would cause the United States to incur any new international obligations or liabilities.
8. It is possible that other developed countries made their various proposals for reductions of carbon dioxide and other greenhouse gases with the knowledge that U.S. opposition to such measures would preclude them from having to fulfill their proposals. It is just as possible, however, that some or all of those countries were seriously interested in reducing greenhouse gas emissions.
9. Philip Shabecoff, *A New Name for Peace: International Environmentalism, Sustainable Development, and Democracy* (Hanover, NH: University Press of New England, 1996), p. 152.
10. Ibid., p. 153.
11. Albert Gore, "U.S. Support for Global Commitment to Sustainable Development," address to the Commission on Sustainable Development, United Nations, New York City, 14 June 1993, *U.S. Department of State Dispatch* 4, 24 (14 June 1993).
12. Robert Ryan, Jr., former Director, U.S. Office of the United Nations Conference on Environment and Development, Department of State, telephone interview by author, 12 January 1996.

13   The change in the U.S. position to support a legally binding agreement was greeted with pleasure by many environmental nongovernmental organizations, some of which predicted that the change in U.S. policy would force many other delegations into supporting a legally binding protocol. "U.S. Shift in Position," *Earth Negotiations Bulletin* 12, 38 (22 July 1996).

14   U.S., House, Joint Hearings, Committee on Foreign Affairs and Committee on Merchant Marines and Fisheries, testimony of Curtis Bohlen, 26 February 1992, *U.S. Policy Toward the United Nations Conference on Environment and Development*, 102nd Cong., 2nd. sess., 26, 27 February, 21, 28 July 1992, p. 16.

15   Determining "incremental" costs has proved to be difficult and controversial.

16   U.S., House of Representatives, Committee on Foreign Affairs, Subcommittee on Human Rights and International Organizations, statement of Gareth Porter, Environmental and Energy Study Institute, 16 September 1991, *U.S. Policy Toward the 1992 United Nations Conference on Environment and Development*, 102nd Cong. [17 April, 24 July, and 3 October 1991], 1992, p. 271.

17   James Baker, "Diplomacy and the Environment," address before the National Governors Association, Washington, DC, 26 February 1990, *U.S. Department of State Dispatch* 1, 1 (3 September 1990).

18   William K. Reilly, "Renewing the Earth: Economics and the Environment," remarks at the UNCED opening session, Rio de Janeiro, Brazil, 3 June 1992, *U.S. Department of State Dispatch* Supplement 3, 4 (July 1992).

19   Alan D. Hecht and Dennis Tirpak, "Framework Agreement on Climate Change: A Scientific and Policy History," *Climatic Change* 29 (1995 [draft of 18 October 1994]), note 47, p. 402.

20   More efficient use of U.S. development resources is increasingly achieved by giving modest sums to developmental NGOs, rather than administering aid programs directly by Agency for International Development personnel or recipient country governments. The Clinton administration planned to channel nearly half of U.S. foreign aid through NGOs by the year 2000. Barbara Crossett, "Gore Says U.S. Will Shift More Foreign Aid to Private Groups," *New York Times*, 13 March 1995, p. A7.

21   Ryan.

22   Gareth Porter and Janet Welsh Brown, *Global Environmental Politics* (Boulder: Westview Press, 1991), p. 129.

23   Reilly, "Renewing the Earth."

24   George Bush, "America's Commitment to the Global Environment," address at Goddard Space Flight Center, Greenbelt, MD, 1 June 1992, *U.S. Department of State Dispatch* Supplement 3, 4 (July 1992).

25   Shabecoff, *A New Name for Peace*, p. 163. In a separate statement attached to the Biodiversity Convention, the United States declared the following: "As a matter of substance, we find particularly unsatisfactory the text's treatment of intellectual property rights; finances, including, importantly, the role of the Global Environment Facility (GEF); technology transfer and biotechnology."

26   U.S., Congress, Senate, *Convention on Biological Diversity: Message from the President of the United States*, 20 November 1993, Treaty Doc. 103-20 (Washington: U.S. Government Printing Office, 1993).

27   See David E. Pitt, "Biological Pact Passes Into Law," *New York Times*, 2 January 1994, p. 9.

28 The developing countries refused to negotiate a forest *treaty* because to do so might be interpreted as acceptance of the developed countries' assertion that forests are a common heritage of humankind.

29 William Reilly, "U.S. Delegation Press Briefing," Rio de Janeiro, Brazil, 8 June 1992, *U.S. Department of State Dispatch* Supplement 3, 4 (July 1992).

30 U.S., *U.S. Policy Toward the 1992 United Nations Conference on Environment and Development*, statement of Gareth Porter, p. 269.

31 The chief U.S. UNCED negotiator, Curtis Bohlen, said as much to Congress: "So at the [third prepcom] meeting in Geneva last week, we put forth new proposals for broadening the governance of the facility to make it more acceptable to developing countries." U.S., *U.S. Policy Toward the United Nations Conference on Environment and Development*, testimony of E.U. Curtis Bohlen, Assistant Secretary of State for Oceans and International Environmental and Scientific Affairs, 26 February 1992, p. 11.

32 Marvin S. Soroos, "From Stockholm to Rio: The Evolution of Global Environmental Governance" in Norman J. Vig and Michael E. Kraft, eds., *Environmental Policy in the 1990s* (Washington: CQ Press, 1994), p. 310.

33 Baker, "Diplomacy and the Environment."

34 George Bush, "U.S. Policy on the Environment and Development," departure remarks, Andrews Air Force Base, Maryland, 11 June 1992, U.S. Department of State Bulletin *Dispatch* Supplement 3, 4 (July 1992).

35 Reilly, "Renewing the Earth."

36 "Statement by the U.S. Delegation on Poverty, Environmental Degradation, Sustainability, Health, and Education," Preparatory Committee for the 1992 UN Conference on Environment and Development, Geneva, 29 August 1991, p. 1.

37 U.S., *U.S. Policy Toward the United Nations Conference on Environment and Development*, statement of E.U. Curtis Bohlen, 21 July 1992, p. 144.

38 Congressional Research Service, *International Environment: Briefing Book on Major Selected Issues*, report prepared for the U.S. House of Representatives Committee on Foreign Affairs, July 1993.

39 Timothy E. Wirth, "Sustainable Development and National Security," address before the National Press Club, Washington, DC, 12 July 1994, *U.S. Department of State Dispatch* 5, 30 (25 July 1994).

40 Warren Christopher, "Overview of 1995 Foreign Policy Agenda and the Clinton Administration's Proposed Budget," statement before the Senate Foreign Relations Committee, Washington, DC, 14 February 1995, *U.S. Department of State Dispatch* 6, 8 (20 February 1995).

41 See Vivian Lowery Derryck, "Sustainable Development: An Integral Component of U.S. Foreign Policy in Africa," Remarks to the African Studies Association Conference, 12 November 1999, <http://www.usaid.gov/regions/afr/speeches/asaspeech_1.html>.

42 For my somewhat extended discussion of U.S. policy toward common but differentiated responsibility, see Paul G. Harris, "International Norms of Responsibility and U.S. Climate Change Policy," in Paul G. Harris, ed., *Climate Change and American Foreign Policy* (New York: St. Martin's Press, 2000), pp. 225-239.

43 United Nations Framework Convention on Climate Change (UN FCCC), 9 May 1992, 31 *International Legal Materials* 849 (1992), Art. 2.

44 See Energy Information Administration, *Emissions of Greenhouse Gases in the United States 1996* (Washington: Energy Information Administration, 1997); Energy Information Administration, *Annual Energy Review 1996* (Washington: Energy Information Administration, 1997). This figure may be even higher given the continued U.S. economic expansion through 2000. In 1998, U.S. greenhouse gas emissions were over 11 percent above those of 1990. See Environmental Protection Agency, *Inventory of U.S. Greenhouse Gas Emissions and Sinks: 1990-1998* (Washington: EPA, 2000), <http://www.epa.gov/globalwarming/publications/emissions/us2000/index.html>.

45 FCCC Conference of the Parties, "Review of the Implementation of the Convention and of Decisions of the First Session of the Conference of the Parties: Ministerial Declaration," Conference of the Parties Second Session, Geneva, 8-19 July 1996 (Geneva: United Nations, July 1996), UN Doc. FCCC/CP/1996/L.17, p. 3.

46 UN FCCC, Preamble, Arts. 3 and 4.

47 Ibid., Preamble, where the Convention notes, *inter alia*, that "the largest share of historical and current global emissions of greenhouse gases has originated in developed countries."

48 Specific commitments to limit greenhouse gas emissions apply to OECD countries (except Mexico, which joined in 1994) and twelve Eastern European and former Soviet "economies in transition." The poorest countries of the world are excluded from commitments, but so too are South Korea, Singapore, Saudi Arabia, and similarly "less developed"—but hardly poor—countries. The United States wanted the CBDR principle to be extended to countries within the non-Annex I (developing countries) group. Under such a formula, the poorest countries would remain exempt from commitments, but the relatively affluent developing countries and the largest greenhouse gas emitters would agree to emissions limitations.

49 UN FCCC, Art 4(2)(a).

50 Berlin Mandate, 6 June 1995, UN Doc. FCCC/CP/1995/7/Add.1.

51 Ibid., Art. (I)(1)(a).

52 Ibid., Art. (I)(1)(d).

53 Ibid., Art. (I)(1)(e).

54 Cf. Group of Seven Industrialized Countries (G-7) and Russia, "Final Communiqué of the Denver Summit of the Eight," Denver, 22 July 1997, paras. 14-17.

55 This was evident in press reports. See, for example, William K. Stevens, "Greenhouse Gas Issue Pits Third World Against Richer Nations," *New York Times*, 30 November 1997.

56 "After Kyoto, New Round of Battle Coming Up," *Journal of the Group of 77* (September/November 1997), <http://www.g77.org/Journal/sepnov97/06.htm>.

57 "Report of the Third Conference of the Parties of the Framework Convention on Climate Change: 1-11 December 1997," Summary Issue of the *Earth Negotiations Bulletin* 12, 76 (13 December 1997).

58 Kyoto Protocol to the United Nations Framework Convention on Climate Change, 10 December 1997, UN Doc. FCCC/CP/1997/L.7/Add.1, Annex B.

59 Ibid., Art. 10. Note that common but differentiated responsibility applies to all countries, not just the developed-developing country relationship. For example, while both the United States and Canada must act before the developing countries, the United States must reduce its emissions more than Canada. But the United States, far and away the largest source of greenhouse gases and the wealthiest economy in the world, is required to reduce its emissions less than the European Union (whose

citizens produce fewer greenhouse gases in aggregate and especially less per capita than do the Americans). Most developed countries are required to reduce their greenhouse gas emissions, but Australia (for example) is permitted an eight percent increase. This differentiation is ostensibly based on national circumstances, but in reality it was largely a function of political bargaining in the Kyoto process.

60 When I refer to the U.S. government here I mean the Clinton administration (and the applicable agencies of the Executive branch) *and* the Congress, notably the Senate, which must ratify international treaties signed by the president.

61 The greatest divergences from the U.S. position are held by many developing countries, notably China, who do not want any treaty references to new developing country commitments. European Union member countries also advocated avoiding such references in the near term.

62 White House, "Comprehensive Framework for Effective, Sensible Action," White House Press Release, 22 October 1997, <http://www.whitehouse.gov/Initiatives/ Climate/framework-plain.html>; White House, "Remarks by President Clinton on Global Climate Change," National Geographic Society, Washington, DC, 22 October 1997, <http://www.whitehouse.gov/Initiatives/Climate/19971022-6127.html>.

63 S.Res. 98, 25 July 1997; the transcript of the Senate floor debate and the Byrd-Hagel Resolution are found in "Expressing Sense of Senate Regarding UN Framework Convention on Climate Change," *Congressional Record*, 25 July 1997, pp. S8113-S8139.

64 *Congressional Record*, 25 July 1997, p. S8138.

65 The following arguments are found in *Congressional Record*, 25 July 1997, passim. pp. S8113-S8139. Senators also expressed their views beyond official forums. See, for example, Jack Kemp, "A Treaty Built on Hot Air," *Wall Street Journal*, 25 July 1997.

66 This is based on the widespread belief, backed by most relevant scientists, that there is a climate change problem. It is important to note, however, that some U.S. legislators in both the Senate and House of Representatives still doubt that climate change is a significant problem (or at least that is what they have said publicly).

67 *Congressional Record*, 25 July 1997, passim. pp. S8113-S8139.

68 Ibid.

69 See *Earth Negotiations Bulletin* 12, 76 (13 December 1997).

70 *Congressional Record*, 25 July 1997, p. S8124.

71 Ibid., p. S8117.

72 Ibid., p. S8120.

73 Ibid., p. S8125.

74 Ibid., p. S8128.

75 Ibid.

76 Ibid., p. S8129.

77 Ibid., p. S8131.

78 White House, "Remarks by President Clinton on Global Climate Change".

79 White House, "President Clinton's Climate Change Proposal," 22 October 1997, http://www.whitehouse.gov/Initiatives/Climate/proposal-plain.html.

80 See White House, "Comprehensive Framework for Effective, Sensible Action."

81 White House, "Remarks by President Clinton on Global Climate Change".

82 Ibid.

83 William Schomberg, "Clinton Says U.S. Ready for Substantial Gas Cuts," Reuters, 14 October 1997.

84 Mark Golden, "Climate Treaty Tough Sell for Clinton at Home and Abroad," *Climate News*, 17 October 1997, <http://listproc.mbnet.mb.ca:8080/guest/archives/CLIMATE-L/climate-l.9710/msg00019.html>.

85 White House, "President Clinton's Climate Change Proposal".

86 Senate Foreign Relations Committee, Subcommittee on International Economic Policy, Export and Trade Promotion, Statement of Timothy Wirth, 105th Cong., 1st sess., 9 October 1997, <http://www.state.gov/www.global/oes>.

87 Ibid.

88 Ibid.

89 Ibid.

90 Ibid. As one solution to the whole question of developing country participation, the Clinton administration proposed that the most affluent, non-Annex I, developing countries (e.g., South Korea, Singapore, Saudi Arabia and the like) be placed in a new "Annex B" with different—and more stringent—commitments than the poor developing countries. See, for example, Timothy Wirth's statement before the House Committee on International Relations, 24 July 1997, <http://www.state.gov/www/global/oes/ 970724tw.html>.

91 Albert Gore, Press Conference at the United Nations Committee on Climate Change, Conference of the Parties, Kyoto, Japan, 8 December 1998, http://www.state.gov/www/global/oes/ 971208a_gore_cop.html.

92 Senate Foreign Relations Committee, Statement of Stuart E. Eizenstat, Under Secretary of State for Economic, Business and Agricultural Affairs, 105th Cong., 2nd sess., 11 February 1998, <http://www.state.gov/www/policy_remarks/1998/980211_eizenstat.html>. Therein Eizenstat said: "Some developing countries believe—wrongly—that the developed world is asking them to limit their capacity to industrialize, reduce poverty and raise their standard of living. . . . In determining what developing countries ought to do, we should be aware that the circumstances of developing countries vary widely, along a kind of continuum. . . . Any 'one-size-fits-all' approach to the 'meaningful participation of developing countries' and to satisfy the Byrd-Hagel Resolution is thus unlikely to prevail. . . . Recognizing our 'common but differentiated responsibilities and respective capabilities,' it will be necessary to develop an approach that provides for a meaningful global response to the threat of global warming, while acknowledging the legitimate aspirations of developing countries to achieve a better life for their peoples. . ."

93 Wirth subsequently left the Clinton administration, some say because he thought the administration's proposals for lowering U.S. greenhouse gas emissions did not go far enough. (This was suggested by press reports, such as: John H. Cushman, Jr. and David E. Sanger, "No Simple Fight: The Forces That Shaped the Clinton Plan," *New York Times*, 28 November 1997.)

94 See *Earth Negotiations Bulletin* 12, 69 (2 December 1997); William K. Stevens, "Talks on Global Warming Open in Kyoto," *New York Times*, 2 December 1997.

95 One U.S. delegate at Kyoto, the Department of State's Daniel Reifsnyder, told reporters at a news conference that "We fully acknowledge that they [the developing countries] are going to grow as their needs for development, you know—as they seek to develop. But what we're looking for, I think, is an effort to try to, if you will, abate the increase in those emissions. . . ." Another delegate (Robert Dixon, Director, U.S. Initiative on Joint Implementation) said that "we are interested in developing countries at least agreeing to a process by which they would come to accept their own binding targets, even though those might be growth targets. . . ." U.S. Delegation to the 3rd

Conference of the Parties, Press Briefing, Kyoto, Japan, 5 December 1997, <http://www.state.gov/www/global/ oes/971205_cop.html>.

96   Ibid. and *Earth Negotiations Bulletin* 12, 76 (13 December 1997).
97   U.S., *U.S. Policy Toward the 1992 United Nations Conference on Environment and Development*, statement of Nicholas A. Robinson, Sierra Club United Nations Representative, 24 July 1991, p. 92.
98   Shabecoff, *A New Name for Peace*, p. 136.

6 International Environmental Equity and U.S. National Interests

What explains the shift toward international environmental equity, especially during the Clinton administration? This chapter suggests that the U.S. government's evolving limited acceptance of international equity as an objective of global environmental policy was in part a function of the imperatives of international cooperation in the environmental field. The nature of the environmental issues themselves was partly determining U.S. policy. The well-being of the United States requires that it act in cooperation with other countries to protect the natural environment. National interests—environmental, economic, and security interests—require that the United States find means to bring about international environmental cooperation. International equity was perceived by the U.S. government as one means to effectuate the needed cooperation. Indeed, other developed countries have similar motivations for joining the emerging consensus on international equity in the environmental issue area.

But this is only part of the explanation for U.S. government's limited acceptance of international equity as an objective of its global environmental policy. As the next two chapters argue, political pressures from various actors and interests, as well as the appeal of international equity as a valuable idea in itself, were also important determinants of U.S. policy.[1]

The Government's Posture on Environment and U.S. National Interests

The U.S. government gradually accepted the notion that environmental changes could threaten U.S. national security, and that it was in America's long-term interest to aid the poor countries in developing in a sustainable fashion in order to prevent or reduce threatening environmental changes and their consequences. There were clear signs in the Bush administration that environmental security was an important U.S. concern, and the Clinton administration embraced the notion of "environmental security" as an important factor in U.S. foreign policy and U.S. defense planning in particular.[2] Bush administration Secretary of State James Baker told the National Governors Association in 1990 that the important American foreign policy goals of democracy, prosperity, security and cooperation were all interconnected with the environment. He said that for this reason he and President Bush were "committed to ensuring that environmental issues are fully integrated into [U.S.] diplomatic efforts. This is the greening of [American] foreign policy."[3] Like other developed countries, the United States was increasingly willing to buy Southern participation in international environmental institutions by aiding the developing countries to protect their own environments and develop in a sustainable fashion, especially insofar as this was necessary to avert global environmental threats that could directly or indirectly affect the United States and its allies. The hope was that in so doing the United States would also help developing countries become markets for goods and services produced in the United States.

The Bush administration had to contend with a Congress sympathetic to U.S. policies geared toward addressing environmental threats to U.S. interests. In 1989 then Senator Al Gore called for a "sacred agenda" in international relations as part of new requirements for collective security: "policies that enable the rescue of the global environment."[4] In a 1990 speech before the Senate, Democratic Senator Sam Nunn, then chairman of the Senate Armed Services Committee, described the threats to United States national security, including what he called a "new and different threat" to U.S. national security: "the destruction of our environment." Nunn called on the intelligence and defense agencies of the U.S. government to pool their resources to address environmental dangers threatening both the United States and the world.[5] According to Nunn, "The defense establishment has a clear stake in countering this growing threat. I believe that one of our key national security objectives must be to reverse the accelerating pace of environmental destruction around the world. . . . America must lead the way in marshaling a global response to the problem of environmental degradation, and the defense establishment should play an important role."[6]

Similarly, Republican Representative Benjamin Gilman (who would become chairman of the House Committee on International Relations after the Republican sweep of Congress in 1994) reminded his colleagues in early 1992 that CIA director William Gates had emphasized the relationship between environment and national security in House testimony. According to Gilman, "I do not think we can emphasize enough the growing understanding of the intricate and vital relationship between the environment and our surrounding ecosystems and food supply and our security."[7]

In November 1991, NATO added economic, social and environmental problems to its list of major threats to the Alliance.[8] It is unlikely that it could have done so without the approval, and more likely active support, of the U.S. government, in particular the Department of Defense. Subsequently environmental issues became much more prominent considerations in NATO planning.[9]

The U.S. government was even more conscious of environment/security links during the tenure of the Clinton administration. Documents describing the national security strategy of the United States during the Clinton administration were permeated with references to environmental security generally and environmental threats to U.S. interests in particular.[10] To wit: "Not all security risks are military in nature. Transnational phenomena such as terrorism, narcotics trafficking, environmental degradation, rapid population growth and refugee flows also have security implications for both the present and long term American foreign policy. In addition, an emerging class of transnational environmental issues are increasingly affecting international stability and consequently will present new challenges to U.S. strategy."[11]

Timothy Wirth, under secretary of state for global affairs, told an audience in 1994 that "the life support systems of the entire globe are being compromised at a rapid rate—illustrating our interdependence with nature and changing relationships to the planet. Our security as Americans is inextricably linked to these trends."[12] Wirth declared that the United States and other countries "are coming to understand the close connections between poverty, the environment, the economy and security. . . . It is time to retool our approach to national security—recognizing that our economic and environmental futures are one and the same."[13] Wirth's views were indicative of the extent to which policymakers were concerned about the effect that environmental change might have on both the U.S. and global economies. Vice-President Al Gore was well known for his concern for the global environment.[14] In 1994 he said that America's enemy was "more subtle than the British fleet. Climate change is the most serious problem our civilization faces, and it has caused enormous damage before in human history."[15] Secretary of State Warren Christopher said that "In carrying out America's foreign policy, we will of course use our diplomacy backed by strong military forces to meet traditional and continuing threats to our security, as well as to meet new threats such as terrorism, weapons proliferation, drug trafficking and international crime. But we must also contend with the vast new danger posed to our national interests by damage to the environment and resulting global and regional instability."[16]

In several speeches, President Clinton described the connections he saw between environment and national security. In 1993 he said that Americans,

> face the extinction of untold numbers of species that might support our livelihoods and provide medication to save our lives. Unless we act now, we face a future in which the sun may scorch us, not warm us; where the change of season may take on a dreadful new meaning; and where our children's children will inherit a planet far less hospitable than the world in which we came of age. . . . In an era of global economics, global epidemics and global environmental hazards, a central challenge of our time is to promote our national interest in the context of its connectedness with the rest of the world. We share our atmosphere, our planet, our destiny with all the peoples of this world. And the policies I outline today will protect all of us because that is the only way we can protect any of us. . . .[17]

In a speech to the National Academy of Sciences in 1994, the president said that he was influenced by scholarly work on environmental security.[18] He then went on: "If you look at the landscape of the future and you say, we have to strengthen the families of the globe; we have to encourage equitable and strong growth; we have to provide basic health care; we have to stop AIDS from spreading; we have to develop water supplies and improve agricultural yields and stem the flow of refugees and protect the environment, and on and on and on—it gives you a headache."[19] A headache indeed, but he and his administration were taking the issue seriously.

Concern that environmental changes were threatening U.S. national security was not exclusive to Democratic politicians. Echoing his earlier concerns, in 1993 Representative Gilman proposed a bill establishing a National Committee on the Environment and National Security to "study the role in United States national security of security against global environmental threats, in light of recent global political changes and the rise of new environmental threats to the earth's natural resources and life support systems. . . ." The bill stated that (1) new threats to the global environment, including to the earth's climate system, the ozone layer, biological diversity, soils, oceans, and fresh-water resources, had risen in recent years; (2) such threats to the global environment might adversely affect the health, livelihoods, and physical well-being of Americans, the stability of many societies, and international peace; (3) the definition of national security had broadened to include economic security as well as environmental security; (4) with the end of the cold war and the urgency of reversing global environmental degradation recognized at the Earth Summit, the global environment has taken on greater importance to the United States.[20]

Recognition of environmental threats to U.S. national security also became part of the Department of Defense's strategic planning. A 1994 report issued by the Army War College's Strategic Studies Institute said that the DOD ought to proactively address environmental change issues because "the change in the international arena since the end of the cold war has given rise to an entirely new approach to viewing U.S. security interests, and a recognition of environmental factors in international stability and the onset of conflict."[21] According to Kent Butts, the U.S. national security system recognized environmental problems as security threats, and the "term environmental security reflects the [U.S.] national policymaker's view of current threats to U.S. security."[22] Citing the U.S. National Security Strategy documents beginning in 1991, Butts points out that after the cold war these documents changed to reflect the waning of the strategic nuclear threat and the ascendance of regional, economic, and environmental threats to U.S. national interests.[23] The focus of U.S. national security strategy shifted toward regional conflict, protecting the economy as a vital national interest, and addressing environmental issues that threaten U.S. interests. Beginning in 1991, all National Security Strategy documents cited environmental change issues as important components of U.S. national interests.[24]

According to public opinion polls, many American citizens were also concerned that resource scarcity, environmental degradation, and mass migrations would affect their quality of life. Many Americans saw the interests of regions of the world as connected, especially regarding economics, population, and environment. They believed that a growing global population would have an adverse impact on the global environment and their own lives. Of those responding to a 1994 poll, 52 percent said the growth of the world's population would worsen their quality of life and 73 percent said that it would have an adverse impact on the global environment. Fifty-four percent thought that if Third World countries become strong economically, their actions would do less harm to U.S.'s environment. Americans also supported foreign assistance that would protect the environment. While they perceived nuclear proliferation as the first greatest threat to U.S. security, 59 percent ranked "loss of rain forests and their animal or plant species" as the second greatest threat and 56 percent thought "the loss of ozone in the earth's atmosphere" was the third greatest threat.[25] Also in a 1994 poll, 58 percent of the public and 49 percent of the "leaders" polled thought improving the global environment should be a very important foreign policy goal of the United States.[26]

Environmental Change, U.S. National Interests, and International Equity

The U.S. government's recognition of environmental changes as potential threats to U.S. national interests was in part an outgrowth of reports by scholars and scientists arguing that "environmental security" was a function of the connected issues of environment and development throughout the world, especially the developing world.[27] Perhaps the most common explanation for why the U.S. government acceded to (or did not more aggressively oppose) many of the demands of the developing countries for consideration of international equity in the context of UNCED agreements and conventions was, very simply, that the South had something the United States needed: the ability to make or break efforts to protect the global environment on which the United States and its allies rely for their prosperity and security. International environmental equity was seen by the United States and other governments as a way to persuade the developing countries to join international efforts to protect the environment, and as a way to help them develop as markets for U.S. exports. Thus the United States would benefit on several fronts: It might be spared many of the indirect adverse economic consequences of environmental change (such as lost markets in the South); it might be less likely to have to contend with the potential violent conflict that could arise from environmental destruction in developing regions; and Americans might avoid potential threats to their health and well-being.

Threats to the U.S. Economy

In the 1970s and 1980s eminent international commissions made up of scholars, statespersons, and scientists reported that if poverty in developing countries were not addressed the environmental consequences of that poverty would undermine the prosperity of the developed industrialized countries.[28] The Brundtland Commission reported that "Those who are poor and hungry will often destroy their immediate environment in order to survive: They will cut down forests; their livestock will overgraze grasslands; they will overuse marginal land; and in growing numbers they will crowd into congested cities. The cumulative effect of these changes is so far-reaching as to make poverty itself a major global scourge."[29]

Developing countries are a large and growing market for exports of the United States and other developed countries. Developing countries accounted for over one-third of U.S. exports in recent years, helping to make up for trade deficits with Japan and other countries. More than one-half of U.S. agricultural exports go to developing countries, and about four million jobs in the United States are the result of exports to developing countries.[30] (In contrast, economic troubles in the developing countries during the 1980s cost 1.8 million jobs in the United States, according to one estimate.[31]) People mired in poverty and environmentally degraded areas will not become buyers of value-added exports of the United States and other developed countries. Environmental damage can restrain the ability of developing countries to afford even basic commodity exports, let alone services and manufactured products. In addition to bolstering these markets in general, equity provisions in the UNCED agreements and conventions could help in creating opportunities for firms in the North producing new "sustainable" technologies. Countries developing these technologies (e.g., Germany and Japan) were among those most forthcoming at Rio, and the Clinton administration made the export of American environmental technologies a goal of U.S. foreign economic policy.

Lack of economic development in the South coupled with environmental degradation could lead to more people migrating—legally and increasingly illegally—from the poor South to the affluent North.[32] Several scholars have suggested that environmental destruction in poor countries has and is likely to directly or indirectly cause large and increasingly frequent movements of "environmental" refugees.[33] The foreign policy community has been increasingly concerned that population growth could contribute to scarcities of important resources in the developing countries. Those scarcities could lead to violent conflict and environmental refugees that could in turn lead to regional destabilization and greater economic hardship. Environmental degradation, war and natural disasters push people across international borders. Already, according to Weiner, large movements of refugees have resulted from desertification, floods, toxic wastes and the threat of sea level rise.[34] Mintzer and Leonard predict that if climate change causes an increase in the frequency of large storms and stimulates the movement of large numbers of economic and environmental refugees, the coping capacity of existing national institutions could be quickly surpassed.[35] According to Weiner, "Migration and refugee issues, no longer the sole concern of ministries of labor or of immigration, are now matters of high international politics, engaging the attention of heads of states, cabinets, and key ministries involved in defense, internal security, and external relations."[36]

Threats of Violent Conflict

Destruction of the environment undermines the natural systems that support human life and civilization, thereby imperiling countries' "most fundamental aspect of security."[37] Environmental changes and resource scarcities can cause social conflicts that are greater threats to the security of many countries than are traditional military threats.[38] In the waning years of the cold war a new security paradigm slowly developed among the industrialized countries, one that viewed environmental factors as an increasingly important consideration in assessments of "comprehensive security."[39] The UN Conference on Environment and Development showed, according to Russell Frye, that matters of environmental protection have become geopolitical issues.[40] Much international attention during the UNCED process had shifted toward environmental protection, sometimes exceeding attention paid to issues like nuclear disarmament. The combined threats "posed by the poor Third World masses and the global environmental problems meant that a security syndrome was being created: the global environmental crisis became, at least for the North, a security issue. . . ."[41]

A multi-year project of the American Academy of Arts and Sciences directed by Thomas Homer-Dixon seems to confirm that adverse environmental changes can contribute to or possibly cause violent conflict. While most of this conflict is expected to be subnational, it may be persistent, increasing in frequency as environmental resource scarcities become more pronounced in the coming decades. Homer-Dixon cites as the most pressing short-term problems the growing scarcities of cropland, water, forests, and fish. He believes that climate change will not be a major contributor to violent conflict for several decades, and then not the sole cause but rather a force that exacerbates existing resource scarcities.[42] According to Homer-Dixon, the "insidious and cumulative" impacts of environmental scarcity can contribute to population migrations, economic decline, weakening of political regimes, and other problems. These problems, and the violence that can result from adverse environmental changes, will have "serious repercussions" for the security interests of both North and South. For example, China may fragment as a response to environmental destruction, with all the potential attendant consequences that the breakup of such a large country could entail.[43] In addition to the direct consequences of environmental scarcities, these changes "will also hinder countries from effectively negotiating and implementing international agreements on collective security, global environmental protection, and other matters."[44]

Water scarcities may be one of the greatest threats to developed countries' interests in many developing regions of the world. Conflicts over renewable resources, especially river water, may lead to violent confrontations between groups and between countries. Fresh water scarcities are palpable in the Middle East, having already contributed to interstate conflict there.[45] The security of many of the developed countries—the United States, Western Europe, and Japan—is tied in varying degrees to the Middle East. Mintzer and Leonard believe that "if global warming caused a reduction in precipitation and available surface runoff in any one of several international river systems that form the boundary between competing riparian states (e.g., in the Jordan-Litani river system in the Middle East), the resulting shortfall could increase the likelihood of armed conflict in an already troubled region."[46] According to some observers, the next war in the Middle East is not likely to be over oil but rather over scarce water resources made scarcer by environmental degradation.[47] Echoing these scholarly observations, in 1996 Secretary of State Warren Christopher said that the U.S. government would "confront pollution and the scarcity of resources in key areas where they dramatically increase tensions within and among nations. Nowhere is this more evident than in the Middle East, where the struggle for water has a direct impact on security and stability."[48]

Threats to the Health and Well-Being of Americans

Recent climate change reports from the Intergovernmental Panel on Climate Change (IPCC) and the World Health Organization (WHO) have added urgency to the climate change issue. The second assessment report of the IPCC concluded that climate change is underway, that it is at least partly the result of human activities, and that urgent action is required to prevent substantial increases in the earth's temperature during the twenty-first century.[49] The WHO report describes expected consequences to human health from climate change. According to that report, it is now recognized by scientists that climate change, by altering local weather patterns and disturbing life supporting natural systems and processes, will "affect the health of human populations. The range of health effects would be diverse, often unpredictable in magnitude, and sometimes slow to emerge. Adverse effects are likely to outweigh beneficial effects substantially."[50]

The World Health Organization concluded: by the middle of the this century many major cities around the world could be experiencing up to several thousand extra heat-related deaths each year (perhaps not coincidentally, the United States experienced several deadly heat waves during the 1990s[51]); climate change may soon increase substantially the proportion of the world's population living in malaria transmission zones (including Southern portions of the United States); there will be increased risk of infectious disease, asthma and other acute and chronic respiratory disorders and deaths; the world will experience an overall decrease in world cereal production (raising prices and perhaps reducing supplies in the United States[52]); changes in ocean temperatures, currents, nutrient flows and winds may adversely affect aquatic productivity; regional increases in frequency of droughts and heavy precipitation (leading to flooding) will occur, resulting in greater risk of death, injury and starvation (especially in the developing countries) along with widespread incidence of psychological and social disorders; sea level rise will lead to population displacement, loss of agricultural land and fisheries, freshwater salinization, and social disruption, all of which could adversely affect health; coastal storm surges and resulting damage to coastal infrastructure (e.g., waste-water, sanitation, housing, roads) will become more frequent; and increased exposure to UV radiation as a consequence of ozone loss exacerbated by climate change will lead to skin cancers and cataracts.[53]

International Equity as a Way to Promote Environmental Security

One of the greatest potential environmental threats to U.S. interests is climate change. As argued previously, without the cooperation of the developing countries, international efforts to limit climate change and other forms of global environmental change (e.g., stratospheric ozone depletion) will not be successful. It was well known by the early 1990s that with only moderate economic growth in the developing countries, their carbon dioxide emissions would soon exceed those of the North, giving a substantial push to climate change.[54] Indeed, the U.S. Department of Energy reported in 1994 that the carbon emissions of developing countries grew by 82 percent from 1970 to 1992, surpassing the OECD's in 1983 and remaining above 50 percent of global emissions.[55] China by itself, given its rapid economic growth and vast supplies of coal to fuel that growth, could account for 40 percent of global emissions by 2050.[56] It is already the second largest emitter of greenhouse gases, and in coming decades it will surpass the United States. Other large, rapidly developing countries are increasing their carbon dioxide and other greenhouse gas emissions at high rates. In the meantime, the developed countries' emissions will grow slowly or, in several cases, gradually decline per unit of production and per capita, and in some cases even in aggregate. Nevertheless, the growth in the developing countries' greenhouse gas emissions will overwhelm politically feasible near and medium-term reductions by the developed countries.

In its report, the WHO referred to the importance of considering international equity in efforts to protect the global climate:

> Although the adverse consequences of environmentally insensitive economic growth are now understood, wealthy countries cannot expect poorer nations to unilaterally forego the short-term profits to be obtained from use of their natural resources. Economically sustainable development will only be possible if environmentally sound technology is transferred from industrialized to developing countries. If less energy-intensive, affordable technologies are promoted and transferred to developing countries, pollutant emissions will be reduced and the global community as a whole will benefit. If not, poorer nations will have no financial incentives to refrain from using cheap energy-inefficient technology or from harvesting their natural resources.[57]

On the heels of the 1996 IPCC and WHO reports, the Clinton administration announced a new policy on the most difficult burdens of climate change (at least for the United States). Under Secretary of State Timothy Wirth said at the second conference of the parties to the Climate Change Convention held in Geneva during July 1996 that the United States would support a legally binding international agreement, to be agreed by the end of 1997, to reduce greenhouse gas emissions. The United States had previously blocked international efforts to create such a binding agreement. Wirth's statement was the first time the United States came out in support of a protocol that would require it and other countries to reduce their greenhouse gas emissions. At the same time, Wirth defended the IPCC's second assessment report. He said that "human beings are altering the earth's natural climate system," adding that "Human health is at risk from projected increases in the spread of diseases like malaria, yellow fever and cholera, . . . food security is threatened, . . . water resources are expected to be increasingly stressed, . . . coastal areas are at risk from sea level rise." He said that the United States was committed to ensuring that all countries take steps to limit emissions consistent with the Climate Change Convention.[58]

As the previous chapters showed, considerations of international equity became major provisions of international environmental agreements in the period between the 1972 Stockholm Conference on the Human Environment and the 1992 Rio Conference on Environment and Development. This trend is in large measure the result of the South's newfound bargaining leverage in this issue area. As suggested above, it is not possible for there to be an effective agreement on climate change, ozone depletion, and other environmental commons issues without the participation of the large developing countries. China already ranks (in aggregate, but not per capita) among the largest emitters of carbon dioxide and other greenhouse gases. By producing and using chlorofluorocarbons, China and India could cancel out efforts by the North to protect the stratospheric ozone layer. Developing countries with large tropical forests could choose to exploit those forests, thus destroying their biological diversity and their capacity for carbon absorption. "Faced with this prospect," Oran Young observed, "northerners will ignore the demands of the South regarding climate change at their peril."[59]

Just prior to the Earth Summit, Young prophesied that,

> any global bargain about climate change must be accepted by all the major parties as equitable. In the short run, this is likely to result in the establishment of a climate regime that makes explicit provisions for technology transfers, technical training, and additional financial assistance as means of encouraging the developing countries to participate and of assisting those willing to take part to make the economic changes needed to modernize in a way that produces lower levels of greenhouse gas emissions. In the longer term, these initiatives could set in motion a train of events leading to more constructive approaches to the overarching issues of North/South equity that constitute one of the central concerns of international society today.[60]

Indeed, the 1997 Kyoto Protocol was a move by the United States and other developed countries in this direction. The developing countries were excluded from the mandated emissions cuts, thus removing short-term threats to their perceived national interests. Furthermore, the Kyoto conference established the Clean Development Mechanism, which would allow developed countries to fulfill some of their emissions reductions by paying for less costly reductions in developing countries. This would make reductions easier for the developed countries, of course, but it also gave developing countries an incentive to support the climate change regime because they could expect new money for clean economic development.[61] In short, the only way to get an agreement of any kind was to treat the developing countries differently—to treat them more equitably while bearing in mind the developed countries' wealth and their past and ongoing pollution of the atmosphere.

Conclusion

Concerns by the U.S. government about environmental security and its response to public perceptions about the connection between national interests and environmental changes contributed to willingness on the part of the Bush and especially Clinton administrations to consider international equity in the context of the UNCED deliberations and subsequent negotiations. The myriad local, regional and global environmental problems pose threats to the interests of the United States and other developed countries, as well as the developing countries, stimulating in those countries an interest in finding ways like the equity provisions of international environmental treaties (e.g., the Montreal Protocol as amended at London and the UNCED conventions) to promote environmental protection. In the long term climate change may be viewed as among the greatest threats to national security. As a result, as Shue points out, "Even from negotiations that are nothing but narrow rational bargaining, the poor nations are likely to receive substantial transfers to avoid future problems concerning climate."[62]

Provisions for international equity are one way to get more effective international environmental agreements to address environmental problems before those problems lead to conflict, economic stagnation, and movements of environmental refugees. Homer-Dixon suggests that to prevent the turmoil and violence associated with environmental scarcities "rich and poor countries alike must cooperate to restrain population growth, to implement a more equitable distribution of wealth within and among their societies, and to provide for sustainable development."[63] The U.S. government and other governments involved in UNCED and follow-on negotiations seem to have taken Homer-Dixon's recommendations to heart.[64] At least they were important security considerations, which they were not in past decades. Promoting the national interest increasingly meant finding ways to provide for international equity in environmental agreements.

The U.S. government's gradual and partial acceptance of international equity as an objective of global environmental policy, as well as considerations of international equity in the UNCED agreements and conventions, can be explained in part by the U.S. government's desire to promote basic national economic and security interests. The U.S. government sometimes actively promoted or acquiesced to international equity considerations because they were an effective way to get developing countries to participate in efforts to promote the environmental security and national interests of the United States. The United States joined other northern governments in choosing equity as one means to promote their collective national interests. The alternative—to use coercion against the developing countries—would have been unacceptable and unworkable. New and additional funds, technology transfers on preferential terms, and more "democratic" decision making procedures in international environmental financial institutions were part of a North-South bargain for mutual self-interest. These kinds of bargains remain an important part of ongoing UNCED follow-on negotiations, especially with regard to climate change and other environmental issues with global consequences.

Thus, the nature of environmental issues themselves were one determinant of U.S. policy. Many environmental changes threaten U.S. interests. They cannot be effectively addressed without international cooperation, including the cooperation of the developing countries. This structure of environmental issues became apparent during the Bush administration and had some affect on policy. (As the next chapter suggests, however, powerful forces, most notably White House chief of staff John Sununu and like-minded officials, were able to block more forthright policy.) During the Clinton administration environmental security became a widely accepted concept in U.S. foreign policy and in the defense establishment (in part because the concept was shepherded along by Vice President Gore and Clinton administration appointees chosen by him, as the next chapter argues). By the advent of Clinton, and more so during his administration, there was within the U.S. government substantial concern about how to bring about international cooperation to address many environmental change issues that may threaten U.S. interests, and thus much more interest in international equity as a subject of global environmental policy.

Notes

1   Explanations for U.S. international environmental policies are examined in much greater detail in Paul G. Harris, ed., *The Environment, International Relations, and U.S. Foreign Policy* (Washington: Georgetown University Press, 2001) and Paul G. Harris, ed., *Climate Change and American Foreign Policy* (New York: St. Martin's Press, 2000).

2   See, for example, White House, *1996 U.S. National Security Strategy of Engagement and Enlargement* (Washington, DC: The White House, January 1996). The Clinton administration's elevation of environment as a security concern is summarized in Gareth Porter, "Advancing Environmental Security through 'Integrated Security Resource Planning,'" *Environmental Change and Security Project Report* 2 (Spring 1996), pp. 35-38.

3 He went on to say that "together, the earth's peoples must work, so that this precious web of life shall embrace, in beauty and in peace, all the generations to come." U.S., Department of State, "Diplomacy for the Environment" *Current Policy Series* 1254 (1990), quotes are from pp. 2 and 4. This was not a new attitude even then. During the Reagan administration, Deputy Assistant Secretary for Environment, Health, and Natural Resources in the State Department, Richard Benedick, said that the administration had recognized that the national interests of the United States can be undermined by instability in other countries related to environmental degradation, population pressures and resource scarcity. See U.S. Department of State, "Environment in the Foreign Policy Agenda," *Current Policy Series* 816 (1986).

4 Albert Gore, paper presented to the Forum on Global Change and Our Common Future, National Academy of Sciences, Washington, DC, 1 May 1989. In his book, *Earth in the Balance* (New York: Houghton Mifflin, 1992), p. 29, Gore described global environmental changes as "fundamentally strategic."

5 It is worth noting here that this is precisely what happened in the Clinton administration. On the urging of Vice President Gore, the Department of Defense and other executive agencies shared data collected during the cold war (e.g., from satellite imagery and ocean surveillance systems) with scientists. There were even programs to share nearly real-time data with environmental and earth scientists to help them better understand global environmental changes.

6 Sam Nunn, "Strategic Environmental Research Program," speech before the United States Senate, Washington, 28 June 1990.

7 U.S., *U.S. Policy Toward the United Nations Conference on Environment and Development*, 102nd Cong., 2nd. sess., 26, 27 February, 21, 28 July 1992, statement of 26 February 1992, p. 14.

8 See NATO Press Service, "The Alliance's New Strategic Concept," *Press Communiqué*, 7 November 1991.

9 See NATO Committee on the Challenges of Modern Society, "The Challenges of Modern Society: Environmental Clearing House System," <http://www.nato.int/ccms/home.htm>.

10 See, for example, White House, *1995 National Security Strategy of Engagement and Enlargement*, February 1995, p. 21. See also subsequent versions under the same title.

11 White House, *1995 National Security Strategy of Engagement and Enlargement*, p. 1.

12 Cited in Alex de Sherbinin, "World Population Growth and U.S. National Security," *Environmental Change and Security Project Report* 1 (Spring 1995), p. 29.

13 Timothy Wirth, "Under Secretary Wirth's Address Before the National Press Club, 'Sustainable Development: A Progress Report,'" 12 July 1994, p. 9, cited in *Environmental Change and Security Project Report* 1 (Spring 1995), pp. 55.

14 See Gore, *Earth in the Balance*.

15 Albert Gore, "Vice President Gore's Remarks at the White House Conference on Climate Action," 21 April 1994, cited in *Environmental Change and Security Project Report* 1 (Spring 1995), p. 52.

16 Warren Christopher, "American Diplomacy and the Global Environmental Challenges of the 21st Century," address at Stanford University, 9 April 1996.

17 William J. Clinton, "Remarks on Earth Day 1993, 21 April 1993", cited in *Environmental Change and Security Project Report* 1 (Spring 1995), pp. 50-51.

18 President Clinton specifically cited Thomas Homer-Dixon. See, for example, Homer-Dixon, "Environmental Change and Violent Conflict," *Scientific American* (February 1993), pp. 38-45.

19   William J. Clinton, "President Clinton's Remarks to the National Academy of Sciences," 29 June 1994, p. 2, cited in *Environmental Change and Security Project Report* 1 (Spring 1995), p. 51. Note how Clinton combines considerations of environmental security and international equity. More recently, in talks with China's President Jiang Zemin, President Clinton told Jiang that "The greatest threat to our security that you present is that all of your people will want to get rich in exactly the same way we got rich. And unless we try to triple the automobile mileage and to reduce greenhouse gas emissions, if you all get rich in that way we won't be breathing very well. There are just so many more of you than there are of us, and if you behave exactly the same way we do, you will do irrevocable damage to the global environment. And it will be partly our fault, because we got there first and we should be able to figure out how to help you solve this problem. . . . It's one thing, Mr. President, I hope we will be cooperating on in the years ahead, because I think that other countries will support your development more if they don't feel threatened by the environment." Quoted in Thomas Friedman, "Gardening With Beijing," *New York Times*, 17 April 1996, p. A23. Clinton was personally engaged in international climate change negotiations, raising the issue with every head of state he met (according to his last climate change envoy, Frank Loy). See Paul Brown, "Forests: U.S. Concession Fails to Please Experts," *The Guardian*, 21 November 2000. The president called other leaders to discuss the topic during the last conference of the parties to the climate change convention before he left office. William Drozdiak, "Sharp Disputes Snag Global-Warming Agreement at Hague Conference," *Washington Post*, 25 November 2000.

20   United States, Congress, House, *H.R. 575 to Establish the National Committee on the Environment and National Security*, 103rd Cong., 26 January 1993.

21   Kent Hughes, ed., *Environmental Security: A DOD Partnership for Peace*, Strategic Studies Institute Special Report (Washington, DC: U.S. Army War College, 1994), pp. 1-2.

22   Kent Butts, "National Security, the Environment and DOD," *Environmental Change and Security Project Report* 2 (Spring 1996), p. 22.

23   Ibid, p. 24.

24   See *National Security Strategy of the United States* (Washington, DC: GPO, 1991, 1993, 1994). For detailed discussions of environmental security in the context of U.S. foreign policy, see Brad Allenby, "New Priorities in American Foreign Policy: Defining and Implementing Environmental Security" and Jon Barnett, "Environmental Security and American Foreign Policy: A Critical Examination" in Harris, *The Environment, International Relations, and U.S. Foreign Policy*.

25   "Highlights from a Review of Existing Survey Data Regarding American Views on U.S. Leadership and Foreign Assistance, Summary Findings," *Polls and Public Opinion: The Myth of Opposition to Foreign Assistance*, May 1994, <gopher://gaia.infousaid.gov/0r0-13682-/agency_wide/why_foreign_aid/pol>.

26   John E. Reilly, "The Public Mood at Mid-Decade," *Foreign Policy* (Spring 1995), p. 82.

27   The arguments of experts described here suggest the context in which policymakers operated. Some scholars and scientists clearly were involved in the policy process. For example, several experts testified before Congress (e.g., Janet Welsh Brown and Gareth Porter), and the scholarly writings of others were cited by policymakers (e.g., President Clinton's citing of Thomas Homer-Dixon's work on environmental security).

28 Donella Meadows et al. (Club of Rome), *Limits to Growth* (New York: Universe Books, 1972); Brandt Commission, *North/South: A Programme for Survival* (Cambridge: MIT Press, 1980); U.S. Council on Environmental Quality and Department of State, *Global 2000 Report to the President of the United States* (Washington: U.S. Government Printing Office, 1980); World Commission on Environment and Development (Brundtland Commission), *Our Common Future* (Oxford: Oxford University Press, 1987).

29 Brundtland Commission, p. 28.

30 Department of State, "Focus on Diplomacy: The State Department at Work," 19 September 1996, <http://www.state.gov/www/about_state/diplomacy.html>. See also U.S., House of Representatives, Committee on Foreign Affairs, Subcommittee on Western Hemisphere Affairs, statement of Janet Welsh Brown, Senior Associate, World Resources Institute, *The United Nations Conference on Environment and Development*, 102nd Cong. (4 February 1992), 1993, p. 47.

31 U.S., *The United Nations Conference on Environment and Development*, statement of Janet Welsh Brown, p. 47.

32 Norman Myers, *Environmental Exodus: An Emergent Crisis in the Global Arena* (Washington: Climate Institute, 1995).

33 See the work of Homer-Dixon cited throughout this chapter and, for example, Jessica T. Mathews, "The Road From Rio: Glimpses of a New World," paper presented at the From Rio to the Capitals Conference, Louisville, 27 May 1993; Sanjoy Hazarika, "Bangladesh and Assam: Land Pressures, Migration, and Ethnic Conflict," *Occasional Paper Series of the Project on Environmental Change and Acute Conflict* 3 (March 1993), pp. 45-65 and Astri Suhrke, "Pressure Points: Environmental Degradation, Migration and Conflict," in the same paper, pp. 4-43. While Suhrke found connections between environmental change and refugee movements, he determined that such connections are usually indirect and that environment is but one of many causes. In a 1994 poll, 72 percent of the American public saw large numbers of immigrants and refugees coming into the U.S. as a critical threat. John E. Reilly, "The Public Mood at Mid-Decade," p. 87.

34 Myron Weiner, "Security, Stability, and International Migration," *International Security* 17, 3 (Winter 1992/93), pp. 91-93.

35 Irving Mintzer and J. Amber Leonard, "Visions of a Changing World," in Mintzer and Leonard, *Negotiating Climate Change* (New York: Cambridge University Press, 1994), p. 12.

36 Weiner, "Security, Stability, and International Migration," p. 91.

37 Michael Renner, "National Security: The Economic and Environmental Dimension," *Worldwatch Paper* 89 (May 1989), pp. 29-30.

38 See, for example, Jessica Tuchman Mathews, "Redefining Security," *Foreign Affairs* 67 (1989), pp. 162-177; Norman Myers, "Environment and Security," *Foreign Policy* 74 (Spring 1989), pp. 23-41; Myers, *Ultimate Security: The Environmental Basis of Political Stability* (New York: W.W. Norton, 1993); Odelia Funke, "National Security and the Environment," in Norman J. Vig and Michael E. Kraft, eds., *Environmental Policy in the 1990s* (Washington: CQ Press, 1994), pp. 323-45. Literature on the connection between environment and national security is now widespread. See the extensive bibliography in issues of the Woodrow Wilson Center's *Environmental Change and Security Project Report*; and the work of Homer-Dixon cited in this chapter. The North Atlantic Treaty Organization, like the defense ministries of its member states, has acknowledged the connection between environmental change and

national security by establishing a bureau devoted to environmental security. For a NATO perspective on the Earth Summit see Robert Banks, "The Follow-Up to the Earth Summit," Draft General Report, NATO Public Data Service, mimeo, 22 December 1994, from Internet multiple recipients list <NATODATA@ccl. kuleuven.ac.be>.

39   See *Environmental Security: A Report Contributing to the Concept of Comprehensive Security* (New York: International Peace Research Institute, 1989).

40   Russell S. Frye, "Uncle Sam at UNCED," *Environmental Policy and Law* 22, 5/6 (1992), p. 341.

41   Pratap Chatterjee and Mathias Finger, *The Earth Brokers* (London: Routledge, 1994), p. 142.

42   Thomas Homer-Dixon, "Environmental Scarcities and Violent Conflict: Evidence from Cases," *International Security* 19, 1 (Summer 1994), pp. 5-40; quote is from pp. 39-40.

43   Vaclav Smil, "Environmental Change as a Source of Conflict and Economic Losses in China" and Jack A. Goldstone, "Imminent Political Conflicts Arising from China's Environmental Crisis," *Occasional Paper Series of the Project on Environmental Change and Acute Conflict* 2 (December 1992), pp. 5-39 and 41-58 respectively.

44   Homer-Dixon, "Environmental Scarcities and Violent Conflict," p. 36.

45   J.R. Starr and D.C. Stoll, *U.S. Foreign Policy on Water Resources in the Middle East* (Washington: Center for Strategic and International Studies, 1987). See also Peter H. Gleick, "Water and Conflict," *Occasional Paper Series of the Project on Environmental Change and Acute Conflict* 1 (September 1992), pp. 3-28 and Miriam R. Lowi, "West Bank Water Resources and the Resolution of Conflict in the Middle East" in the same issue, pp. 29-52.

46   Mintzer and Leonard, "Visions of a Changing World," pp. 11-12. See also Peter H. Gleick, "Water and Conflict: Fresh Water Resources and International Security," *International Security* 18, 1 (Summer 1993), p. 111.

47   Alan Cowell, "Hurdle to Peace: Parting the Mideast Waters," *New York Times*, 10 October 1993, p. A1; Katrina S. Rogers, "Rivers of Discontent-Rivers of Peace: Environmental Cooperation and Integration Theory," *International Studies Notes* 20, 2 (Spring 1995), pp. 10-21.

48   Warren Christopher, "American Diplomacy and the Global Environmental Challenges of the 21st Century," address at Stanford University, 9 April 1996.

49   The full IPCC report is contained in *Climate Change 1995*, 3 vols. (New York: Cambridge University Press, 1996).

50   World Health Organization, "Executive Summary," *Climate Change and Human Health* (Geneva: WHO Office of Global Integrated Environmental Health, 1996), p. 2.

51   In July 1998 President Clinton attributed—perhaps prematurely from a scientific standpoint—widespread forest fires then afflicting Florida to climate change. "Clinton Links Fires, Global Warming," *Climate News*, 13 July 1998.

52   "Some mid-continental drying in temperate zones, such as the mid-west USA . . . may occur." World Health Organization press release, "Climate Change and Human Health," WHO Office of Global and Integrated Environmental Health, 12 July 1996, p. 1.

53   Ibid. See also Ross Gelbspan, *The Heat is On: The High Stakes Battle Over Earth's Threatened Climate* (New York: Addison-Wesley, 1997) and Johns Hopkins University, "Climate Change and Human Health: The Climate Change and Human Health Integrated Assessment Website," <http://www.jhu.edu/~climate/>.

54 Thomas Homer-Dixon, "Physical Dimensions of Global Change," in Nazli Choucri, ed., *Global Accord· Environmental Challenges and International Responses* (Cambridge, MA: MIT Press, 1995), p. 62; Vaclav Smil, "Planetary Warming: Realities and Responses," *Population and Development Review* 16, 1 (March 1990), p. 18; Peter Gleick, "The Implications of Global Climatic Change for International Security," *Climate Change* 15 (1989), pp. 309-25. (These citations are given to indicate that such ideas were extant before the Earth Summit.)

55 United States Department of Energy, *Energy Use and Carbon Emissions. Some Interpretations and Comparisons* (Washington, DC: Energy Information Administration, 1994).

56 Nancy C. Wilson, "China Faces Hard Energy Choices: Booming Economy, Soaring Emissions," *Climate Alert* 6, 3 (May-June 1993).

57 WHO press release, p. 3.

58 "U.S. Supports Agreement on CO_2 Reduction at U.N. Climate Conference," *Deutsche Press-Agentur*, 17 July 1996, p. 16 (trans. Environmental Media Services).

59 Oran R. Young, "Negotiating an International Climate Regime: The Institutional Bargaining for Environmental Governance," in Choucri, pp. 431-52; China's greenhouse gas figures are from p. 447. Similarly, Susan George predicted before the Earth Summit that "Biospheric solidarity may yet be forced upon the governments of the North; like it or not, they may finally have to recognize that debt relief and real contributions to sustainable development in the South are vital to their own survival." Susan George, "Managing the Global House: Redefining Economics in a Greenhouse World," in Jeremy Leggett, ed., *Global Warming: The Greenpeace Report* (Oxford: Oxford University Press, 1990), pp. 438-456.

60 Young, p. 447. Young's chapter was prepared before the Earth Summit.

61 For studies of how the Clean Development Mechanism can help the world's poor, see Peter G. Taylor and Ken Fletcher, *Prioritizing Opportunities under the Clean Development Mechanism* (London: Department for International Development, 2000).

62 Henry Shue, "The Unavoidability of Justice," in Andrew Hurrell and Benedict Kingsbury, eds., *The International Politics of the Environment* (Oxford: Clarendon Press, 1992), p. 375.

63 Homer-Dixon et al., "Environmental Change and Violent Conflict," p. 45.

64 As shown above, President Clinton directly cited Homer-Dixon as having dramatically influenced his thinking. See especially President Clinton's address to the National Academy of Sciences, Washington, DC, 29 June 1994, "Advancing a Vision of Sustainable Development," reprinted in *U.S. Department of State Dispatch* 5, 29 (18 July 1994).

7 International Environmental Equity and American Politics

The previous chapter explained the U.S. government's partial acceptance of international equity as an objective of global environmental policy by reference to concerns about the national interests of the United States that derived from the nature of environmental issues. The next chapter examines the role of normative ideas as possible factors in shaping U.S. policy in this regard. This chapter introduces another explanation for U.S. policy toward international environmental equity. What explains the Bush administration's change from radical opposition to international equity provisions in the UNCED deliberations to policies that, while not embracing equity, were closer to the positions of other developed countries and to the demands of the developing countries? What explains the Clinton administration's shift toward acceptance of international equity as an objective of global environmental policy, albeit a very modest one that did not go as far as even Clinton may have wished? Many expected the Clinton administration to be sympathetic toward all environmental issues (more than it actually was), so perhaps the real burden is to explain why it did not go further in this regard, and especially to explain why the Bush administration went further than many would have expected. This chapter argues that the U.S. government's partial acceptance of international equity was in part a result of the demands made by politically active and influential actors, both within and outside the government and within and (to a lesser degree) outside the United States.

Equity considerations have acted as an indirect means by which the U.S. government and individuals therein—as well as similar actors abroad—have promoted their interests; equitable results have come from purely self-interested actions by the U.S. government and individuals in the policy process, apart from concerns for international equity per se. The government was persuaded to "give in" to demands for equity to satisfy the demands of individuals, industry and environmental interest groups, the public and Congress, foreign governments and other actors. Constituencies in the United States (and other developed states) concerned about the development of international institutions for environmental protection and international equity lobbied American politicians and bureaucrats. Similarly, international nongovernmental organizations (NGOs) and foreign diplomats pressured the U.S. government and its delegates involved in the actual UNCED negotiations through moral suasion and threats to expose the injustices of potential agreements. All of these actors operated in the context of the pluralist American political process—including the omnipresent concern of U.S. politicians for reelection and especially the 1992 and 1996 presidential elections—as well as the international context of the emergence of international equity as an objective of international environmental institutions and other governments.

Such pressures resulted in the United States refraining from being even more of a "spoiler" at the Earth Summit, thereby opening the way for more progressive forces to use their influence to increase the prominence of international equity considerations in the UNCED deliberations and subsequent agreements and treaties. This mirrored what occurred during the 1990 London meeting on the Montreal Protocol. The Bush administration accepted equity provisions (including the Multilateral Fund) in the Montreal Protocol's London amendments because it wanted to cater to concerns of industry groups representing American makers of chlorofluorocarbon substitutes and to demands of important foreign actors (e.g., British Prime Minister Margaret Thatcher) that viewed equity considerations as important means to persuade developing countries to join the Montreal Protocol. This occurred despite the Bush administration's desire, in light of then nascent negotiations on climate change, to avoid setting a precedent for North-South funding of environmental agreements. Similar forces acted on the Clinton administration to persuade it to be more responsive to policies that might promote international equity in the context of environmental issues.

Thus one set of reasons for the U.S. government's acceptance of international equity as an objective of global environmental policy were the influences of the many actors involved in the highly pluralistic policy-making process.

Political Forces Acting on the Bush Administration

The most interesting question about the Bush administration's international environmental policies is not "Why was the Bush administration unwilling to go as far as many other developed countries in agreeing to greater consideration of international equity in the UNCED deliberations?," but rather "Why, considering how opposed the U.S. government under President Bush was to greater consideration of equity, did the Bush administration go as far as it did toward the positions of countries more sympathetic to considerations of international equity in international environmental negotiations and the UNCED process?" In resisting calls by other countries to be more forthcoming at Rio, Bush acted in a manner consistent with his past policies. His administration, like President Reagan's, generally opposed new funding and concessional technology transfer to the developing countries, and there was no desire to take responsibility for America's flagrant and disproportionate pollution of the global atmosphere. The United States was brought along kicking and screaming to agreement on (or at least acquiescence to) substantive provisions for international equity in the 1990 London amendments to the Montreal Protocol. It took a variety of forces—including the intervention of other heads of government—to get cooperation from the United States at the London meeting. Likewise at Rio.

President Bush went to Rio with little in the way of new funds for sustainable development or commitments to take on the burdens of cleaning up the global environment. Nevertheless, he did pledge to double U.S. global forest assistance and to assist with what he called an "extensive" program of "technology assistance." He said that he recognized that the developing countries would need assistance for sustainable development, and he pledged to increase U.S. international environmental aid by 66 percent over 1990 levels in addition to the $2.5 billion the United States was already providing for Agenda 21 type programs through international development banks.[1] He pledged that his administration would act to reduce carbon dioxide emissions and push for rapid ratification of the Climate Change Convention (in its now-weakened form). That President Bush went to Rio at all was a surprise to many, considering his anti-environmental rhetoric as the 1992 presidential campaign got underway and his refusal to commit to participation in the conference until almost the last minute.

The presidential election process in the United States was a vital consideration for the Bush administration in its assessments of the UNCED Rio conference and related international deliberations. While President Bush may have had some sympathies toward environmental protection (as suggested by his record before being elected vice president), as the Earth Summit and the elections approached he did not want to appear sympathetic to environmental issues because to do so might jeopardize support among Republican party loyalists and "Reagan Democrats," especially those in electorally-important Southern and Rocky Mountain states, that had contributed to previous Republican presidential victories.[2] Bush was pressured by right-wing Republicans who were increasingly dominant in the party, as well as by isolationist Patrick Buchanan, who was challenging him for the Republican presidential nomination and who opposed environmental regulations. The lagging U.S. economy, along with other domestic problems, contributed to concerns among Republicans that President Bush had devoted too much of his time to international issues. A well-publicized trip to Brazil in which he would give new funds and American technology to the developing countries might have had the makings (from the perspective of the Republicans) of a public relations disaster.

The right wing of the Republican party took much of its guidance from the conservative Heritage Foundation think tank. According to a Heritage Foundation report, U.S. acquiescence to the demands of the developing countries at the Earth Summit could "affect profoundly" economic growth and productivity in the United States, and might reduce America's international competitiveness. The report called on the Bush administration to protect intellectual property rights and to refuse concessional transfers of biotechnologies, to counter all proposals for additional spending on the environment, and to oppose any specific targets for the reduction of greenhouse gases.[3]

Alternatively, looking like a spoiler at Rio could bolster the criticisms of his administration coming from Bush's Democratic opponents.[4] Nevertheless, "President Bush's political advisers apparently concluded that standing alone in opposition to international agreements would play well with his conservative constituency, which did not concede the seriousness of global environmental problems and had exhorted him not to attend a conference dominated by the world's 'environmental extremists.'"[5] In the event, Bush went to the Earth Summit despite calls by fellow Republicans to stay at home, but not without having gotten agreement from other countries that the Climate Change Convention would not commit the United States to specific targets for reductions of greenhouse gases. Several Bush advisors had convinced him that he could not afford to stay away from the Earth Summit because to do so would be an abdication of global leadership and would harm his environmental record among young voters in some large states, including Florida and California.[6] It seems that Bush was trying to simultaneously appease environmental voters by attending the Summit and Republicans by bringing very little with him to it. He attempted to balance growing public interest in the environment and the interests of business, industry, and the Republican right wing. Nevertheless, in refusing to move closer to the position of other industrialized countries on, for example, limits on greenhouse gas emissions and biodiversity, Bush alienated world opinion and the American environmental movement. At the Earth Summit, the United States was isolated from all but a handful of countries.[7]

Members of Congress pressed the Bush administration from many angles, but the Democrat-controlled Congress was generally supportive of the UNCED process, including steps to help the developing countries with sustainable development.[8] Several resolutions were introduced in the Senate and the House of Representatives calling on Bush to attend the Earth Summit and declaring the legislature's strong support for the "strongest possible versions of all the treaties scheduled for discussion."[9] Concurrent Resolution 263 of 1991, which expressed the sense of Congress with regard to U.S. UNCED policy, stated that President Bush ought to give UNCED his personal attention and that the United States should (among many other things): "place the highest priority on the success" of UNCED; propose an initiative on funding Agenda 21 and other global environmental cooperation efforts that "takes into account the concerns of developing countries regarding additional costs of international environmental protection and the basic development goals of those countries"; and support international efforts "aimed at identifying ways that poverty can be alleviated and natural resources better conserved through reduction of developing country debt burdens."[10] What is more, seventy-four members of Congress sent a letter to President Bush urging him to attend the Rio conference, to commit the United States to stabilizing U.S. carbon dioxide emissions at 1990 levels by 2000, and to help establish international institutions with "means to address the interrelated issues of poverty, environmental degradation, development, and population."[11]

Both houses of Congress sent delegates to UNCED deliberations as congressional advisors. Four members of congressional staffs were part of the U.S. delegation beginning with the March 1991 second prepcom in Geneva. Bush faced extensive criticism from congressional members of his own delegation to the conference, especially Democratic Senators Timothy Wirth (subsequently under secretary of state for global affairs in the Clinton administration) and Al Gore (who served as chairman of the Senate delegation). According to a high-level U.S. diplomat involved in UNCED, Gore was probably the single most active person in Congress on UNCED issues. Senator Gore attended prepcom meetings, "he did a lot of homework and gave speeches [and] he was on the phone to us all the time."[12] Throughout the pre-summit UNCED process and during the Earth Summit, a number of Congress members spoke of their dissatisfaction with the diplomatic positions taken by the Bush administration and the American delegation. During the Rio summit, they occasionally took their criticisms to the nearby Global Forum of nongovernmental organizations to convey a different U.S. perspective on the issues being negotiated.[13]

Following the Earth Summit, some members of Congress made statements especially sympathetic to the equity objectives of UNCED. Representative Dennis M. Hertel told his colleagues that it was up to the United States "to contribute energetically to raising the standards of living and the state of environmental awareness in developing countries; to prove that we are unafraid of reducing our own consumption; and to own up to our responsibility to prudently utilize the earth's natural resources."[14] In a statement before a hearing of the House Committee on Foreign Affairs considering U.S. UNCED follow-up, chairman Dante Fascall chose to emphasize the vastness of global poverty. Noting that 1.1 billion people were living in acute poverty, one-third of world's population was without adequate sanitation, and one billion people were without safe drinking water, he said, "alleviating poverty is both morally imperative and essential for environmental sustainability."[15]

Public awareness of climate change increased during the early negotiations on climate change at the end of the 1980s.[16] At the Noordwijk meeting on climate change in November 1989, all other developed countries called for target dates for stabilizing emissions of carbon dioxide and for a "climate fund" to help the developing countries pay for their efforts to combat climate change. The Bush administration responded, after intense domestic criticism of its failure to agree to such measures, with $1 billion for research on climate change, a program to plant one billion trees, and a U.S.-sponsored conference on climate change to be held in April 1990.[17] The conference garnered extensive negative public criticism after powerful White House chief of staff John Sununu overruled the Environmental Protection Agency (EPA) and State Department, and wrote a conference speech for Bush that called only for more research.

As the Earth Summit approached, public opinion was increasingly sympathetic to the conference's broader environmental objectives. Bush faced a dilemma in the Earth Summit. While not wanting to be viewed as surrendering to international environmentalists, he also did not want his administration to be seen as the only "spoiler" at the Rio conference, especially in light of the American public's support for the Earth Summit's objectives.[18] In the United States, the Earth Summit was preceded by nationwide celebrations of Earth Day and quite heavy press coverage. Opinion polls showed that Americans largely disapproved of President Bush's environmental record.[19] There was concern among the public (and its representatives in Congress) that the United States was not acting as a leader in the field of environmental protection.[20] According to a *Wall Street Journal* story, the "environmental president" was in danger of being labeled at the Rio conference as "No. 1 enemy of the Earth."[21]

Domestic economic interests were among the variety of forces that influenced the Bush administration's posture on international equity and other aspects of the Earth Summit agenda. The United States, like many other industrialized countries, was experiencing economic recession, meaning that the U.S. government was not very receptive to new international programs that would require large new expenditures of foreign aid or might limit U.S. competitiveness. As the presidential vote neared, American politics were increasingly focused on balancing the federal budget and reducing the accumulated federal deficit. This general atmosphere of concern about the economy made the Bush administration especially sensitive to commercial interests and their advocates within nongovernmental organizations (e.g., business and industry political action committees and industry-funded think tanks).

Business and industry were most interested in the climate change and biodiversity negotiations because they would, if ratified, become binding on the United States. They made less effort trying to influence negotiators involved in the details of Agenda 21, a document that was always expected to be nonbinding[22] This is not surprising, because business interests concerned about climate change pushed the U.S. government in different directions. Some business groups were opposed to strong environmental measures and concessional sharing of technologies. For example, in the case of the biodiversity negotiations, some corporate lobbyists—and even more so several conservative groups who claimed to represent corporations—feared that a biodiversity treaty which met the demands of the developing countries would reduce the value of intellectual property held by American businesses that had developed biotechnologies, especially if those technologies were made with genetic materials harvested from the South.[23] There was some debate after the Earth Summit as to whether biotechnology and other companies were actually opposed to the treaty. Nevertheless, according to William Reilly, "Certainly elements of that industry convinced the State Department, Vice-President's office and White House that the Convention did threaten them; no companies communicated any contrary message, even privately."[24] (Many U.S. biotechnology companies would later reinterpret the convention and see potential benefits in it.)

Several industries opposed the climate treaty because they thought it would increase the cost of energy in the United States and slow economic growth by requiring them to use less oil and coal. The oil and automobile industries joined forces with sympathetic members of Congress and the executive branch to prevent tighter energy efficiency requirements for American cars, a component central to the environmental movement's strategy to reduce U.S. emissions of carbon dioxide.[25] The U.S. coal industry lobbied the government and sent representatives to the climate negotiations in an effort to prevent measures to limit the emissions of carbon dioxide.[26] Coal interests advised the U.S. delegation (and Russian, Saudi and no doubt other delegations) to the climate change negotiations on how to weaken the climate change convention.[27] The coal and utility industries, represented by organizations like the National Coal Association, the Global Climate Coalition and the Climate Council, aggressively opposed carbon taxes and efforts that would obligate the United States to stabilize or reduce its emissions of carbon dioxide.[28] In the months preceding the Earth Summit, a variety of industries opposing environmental restrictions flooded the White House with letters and communicated with the vice president's Council on Competitiveness, which was managing much of the Bush administration's environmental policy.[29]

Some business groups, however, took a more positive view of UNCED, recognizing that they would be called on by consumers to take greater responsibility for the effects their practices have on the environment, and seeing that meeting some environmental regulations (e.g., those requiring greater energy efficiency) could make them more competitive in the long run.[30] This helps explain their tacit alliance with the developing countries in calling for greater official development assistance to pay for technology transfers and to encourage investments by multinational corporations in those countries. The Business Council for Sustainable Development observed that "little foreign direct investment reaches countries near the bottom of the development ladder because the political and social risks of investment are generally too high. [But ODA] can help governments collaborate with business and industry in delivering technology and management systems for basic infrastructure through long-term business partnerships. It can lower the overall financial risks, and may indirectly provide strong leverage to promote political and social reforms. . . ."[31]

Regarding U.S. obligations to reduce emissions of greenhouse gases to limit climate change, several U.S. power companies (Southern California Edison, Los Angeles Water and Power, and New England Electric System) were not among the industries supporting Bush administration efforts to thwart negotiation of a climate change treaty with targets and timetables. Indeed, they committed themselves to reducing their carbon dioxide emissions 20 percent by 2010.[32] In addition, companies promoting energy efficiency technologies, natural gas, and renewable forms of energy supported international regulation of carbon dioxide emissions.

Public pressure, cultivated in large part by environmental NGOs, was the stimulus for UNCED. The Secretary-General of UNCED, Maurice Strong, pushed for NGO involvement in the entire UNCED process, including the prepcoms where the bulk of documents signed at Rio were negotiated.[33] Two hundred NGOs participated in the second UNCED prepcom, and 500 were involved in the final prepcom. The number of NGOs accredited at the official Rio conference reached 1,420, one-third from developing countries.[34] Because Strong viewed them as potential allies, participation of NGOs from developing countries was facilitated by a travel fund established by the UNCED secretariat.[35] Fifteen national delegations included NGO observers. NGOs were seen to be more knowledgeable about environmental issues than many of the participating governments. There was some resistance from United Nations member states to NGO involvement in the UNCED process, but several developed countries—including the United States—supported their inclusion in the process.[36]

Nongovernmental organizations used two strategies to influence the U.S. government's UNCED policy: (1) lobbying during and between the preparatory committee meetings, and (2) participating in the U.S. delegations to the prepcoms and the Earth Summit.[37] Lobbying included making statements in plenary meetings and working groups, drafting concrete proposals, briefing governmental delegates, and participating directly in some informal negotiating sessions. Much of the lobbying occurred in the hallways because most talks were closed to NGO representatives.[38] Nongovernmental organizations from the United States and other developed countries, particularly those NGOs with the greatest financial and personnel resources, had the most influence on the UNCED negotiations. Nongovernmental organizations from developing countries also participated, and U.S. diplomats heard them.[39] Well-organized and funded Environmental NGOs from the United States, as well as those from Canada and Western Europe, lobbied the United States and other national delegations, and they provided the press with ongoing information about the Earth Summit negotiations.[40] NGOs developed regional lobbying networks to push not only the United States but other national delegations. They coordinated strategy through organizations like the Climate Action Networks set up in Europe, Asia and elsewhere. At the New York preparatory committee meeting two months before the summit, NGOs organized issue-oriented task forces that coordinated their lobbying efforts. As a consequence, their draft proposals were discussed by delegates and their ideas affected the final documents.[41]

Nongovernmental organizations most visible in the biodiversity negotiations included the World Resources Institute (WRI), the World Conservation Union, the World Wildlife Fund, and the Consortium for Action to Protect the Earth (CAPE '92, a coalition of American environmental NGOs[42]). NGOs influencing the climate change negotiations included WRI, the Climate Institute, the Climate Action Network, Greenpeace, and CAPE '92. Environmental NGOs attending the NGO Global Forum in Rio demonstrated against U.S. opposition to much of the Earth Summit agenda. Overall, NGOs had partial influence on the cross-cutting issues of new financing and technology transfer that were crucial to all UNCED agreements.[43] The NGOs had much more influence on the chapters of Agenda 21 dealing with poverty and consumption; in these cases their proposals were adopted in part or in whole by the official negotiators.[44] The large U.S. NGOs were especially successful in getting access to the UNCED secretariat and to members of government delegations from the North. Those NGOs were consulted by official participants and were from time-to-time permitted to participate directly in the UNCED process.[45] Environmental and development NGOs met regularly with U.S. diplomats involved in the UNCED negotiations. They played a "relatively strong role, and they were the ones pushing for more accommodation to the Southern agenda."[46]

In addition to lobbying the negotiations from the outside, individuals from NGOs representing industry, research institutes, environmental organizations, and development groups served on the U.S. and other national delegations. At the first prepcom Canada included an NGO participant on its delegation, making it the first country to do so. By the second prepcom several other, mostly developed, countries followed, among them the United States (which had three NGO representatives advising the delegation). This established a trend for future deliberations, including the Rio summit itself.[47] The U.S. delegation to the third prepcom included Gareth Porter of the Environmental and Energy Study Institute. He was highly critical of the U.S. failure to agree to new funding and institutional mechanisms to address sustainable development, and of the U.S. unwillingness to agree to concrete reductions in emissions of greenhouse gases, especially carbon dioxide.[48] By the fourth prepcom the United States had over twenty NGOs on its delegation. Many of the environmentalists who served on the U.S. delegations during the preparatory process felt that they were able to have an impact on the UNCED process, influencing both official members of the U.S. delegation and delegates from other countries.[49]

In the United States, several environmental NGOs took advantage of the departure of John Sununu from the White House in March 1992. He had strongly opposed U.S. actions to promote international environmental equity (however defined). They met with the new chief of staff, Samuel K. Skinner, to discuss climate change issues.[50] Nongovernmental organizations also met with President Bush and William Reilly, although one of them was skeptical of the extent to which they had significant influence on development and poverty issues.[51] Nevertheless, according to Hatch, proponents of a more cooperative U.S. policy on climate change were able to use a variety of strategies to influence the Bush administration:

> Public pronouncements on administration policy in an effort to increase public pressure (prominent politicians such as Sen. Albert Gore, and such environmental groups as the Sierra Club, Natural Resources Defense Council, World Resources Institute and the Environmental Defense Fund were especially active in this regard); lobbying foreign governments to resist pressures from U.S. officials to soften their positions in the international negotiations (again, Gore and environmental groups were involved in this activity); and congressional hearings to focus public attention on the issue. However, the legislative process provided the greatest opportunity to influence the substance of global warming policy, since congressional approval was required if any action were to be take on the administration's energy proposals.[52]

Representatives of environmental and development NGOs testified before Congress on many occasions, consistently expressing their dismay with the U.S. government's failure to be more forthcoming with regard to new development funding, technology transfer, over-consumption and commitments to reduce greenhouse gas emissions.[53] Environmental groups were not able to have the influence on the Bush administration that they had hoped to have, but they were able to persuade Congress to take a more sympathetic stand and to take specific actions, such as including measures in the 1992 National Energy Policy Act that would promote energy efficiency and conservation and encourage the use of renewable energy sources.[54]

Nongovernmental organizations tried to put pressure on the United States and other governments to give equity as much consideration as economic efficiency traditionally received.[55] Environmental NGOs from developing countries were often more concerned about poverty and local developmental issues than about environmental issues, and when they were concerned primarily with the environment that concern was focused on local issues (e.g., sanitation, fishing rights, land use) that most affected the well-being of the poor in their home countries.[56] Rahman and Roncerel note that throughout the debates of the UN's Intergovernmental Negotiating

Committee (INC) on climate change, environmental NGOs used a variety of creative means to push the agenda of poor people, focusing their debates and pressures on equity issues: "The NGOs have traditionally voiced equity concerns which would otherwise remain unheard, with the importance deserved, in the international political processes. Potential victim states discovered that the NGOs participating in the climate negotiations could be constructive partners and conduits of their concerns to the greater community of nations throughout the UN processes."[57]

The important role that individual scientists and coalitions of scientists ("epistemic communities"[58]) can play in U.S. policy formulation was evident in international environmental negotiations that preceded the Earth Summit.[59] International panels of scientists examining the depletion of the stratospheric ozone layer and the causes and consequences of climate change contributed to expert consensus (if not yet agreement of laypersons in the United States) that human activities were adversely affecting the global environment. Scientists had an impact on policy because environmental issues tend to be very complex, often beyond the understanding of most lay government officials and citizens. Non-experts are often left with little choice, short of no action, but to defer to the views of scientists. Of course, no action is often the option chosen.

The Bush administration justified its refusal to support specific limits on greenhouse gases by pointing out that scientists were not certain about the causes and consequences of climate change.[60] The first assessment report of the Intergovernmental Panel on Climate Change (IPCC) issued in 1990, while suggesting strongly that climate change was indeed underway, did not conclude that such changes were caused by human activity.[61] In addition, there were politically active scientists (often financed by the fossil fuel industries) who lobbied actively in an attempt to disprove that climate change was either unusual or caused by humans. Although these scientists were in the minority, they were joined by Sununu in the White House, who consistently insisted on a U.S. policy advocating only more study of climate change, rather than having the United States spend money to promote sustainable development abroad or undertake policies domestically (namely, reduce emissions of carbon dioxide and other greenhouse gases) that might help prevent climate change.[62]

However, by the time of the Earth Summit there was growing consensus among scientists that increasing concentrations of man-made carbon dioxide might have a tremendously adverse impact on the global environment.[63] The vast majority of scientists were calling for more international efforts to limit greenhouse gas emissions.[64] For example, in early 1990, half the U.S. Nobel Prize recipients and half the members of the

U.S. National Academy of Sciences appealed to the U.S. government to take action to limit U.S. emissions of greenhouse gases.[65] In addition, the scientists of the IPCC were influential in making the participants in the UNCED deliberations aware of the potential dangers of climate change. Critics of the administration's climate change policies were bolstered by a draft IPCC report showing a growing scientific consensus on the dangers of climate change and by subsequent pledges from the British and Japanese governments to stabilize their carbon dioxide emissions at the turn of the century.[66] Thus scientists and diplomats (including, informally, many from the United States) increasingly agreed that action should be taken to address environmental degradation, especially climate change, even if more science would be needed to fully understand these problems. Indeed, the "precautionary principle" contained in UNCED statements and agreements means that "lack of full scientific certainty should not be used as a reason for postponing" measures to address global environmental change.[67]

The most important single determinant of the Bush administration's initial policy on international environmental equity was White House chief of staff Sununu. He was generally skeptical of the United Nations, its subsidiary bodies not dominated by the United States, and most of its Secretariat personnel. He had President Bush's ear and carefully monitored the information that the president received regarding the whole UNCED process. In short, most of what little Bush new about UNCED and global environmental issues came from or through Sununu, who had strong credentials in Bush's mind: he had a technical background as an engineer and he was an experienced political operative whom Bush probably assumed to be authoritative when it came to the intersection between environmental issues and domestic politics. Sununu was opposed to U.S. commitments on environmental issues—from action on ozone depletion and climate change to the need and efficacy of North-South financial and technological assistance to address environmental change.

According to Rafe Pomerance, who was an observer and lobbyist during the climate negotiations (and who took charge of climate change negotiations for the State Department early in the Clinton administration), Sununu was the controlling influence on the U.S. delegation to the climate talks.[68] William Nitze suggests that the United States might have moved closer to the position of the other OECD countries in the climate deliberations had it not been for the inordinate influence of a few individuals in the White House, especially Sununu, and because President Bush chose not to become more involved in the process.[69] According to Philip Shabecoff, an intimate observer of the whole UNCED negotiating process, in Sununu "the right had an aggressive agent strategically placed at the inner core of the White House."[70] Until his departure, Sununu or his deputies in the White House determined U.S. policy on the most sensitive global environmental issues, at least insofar as they cared about those issues, which they ultimately did in the cases of concrete commitments to reduce greenhouses gases, concessional technology transfers (especially their major manifestations in the Biodiversity Convention), and financial commitments to fund sustainable development in the South.

Second only to Sununu's own influence was the influence of those individuals in the White House and executive agencies, most notably the Department of State, who were sympathetic to—and even "appointed" by—Sununu. After Sununu left office amid a minor scandal, his influence lingered because so many other like-minded people remained in the Bush administration. Nevertheless, there were many individuals in the U.S. government sympathetic to the considerations of equity in the UNCED deliberations. After Sununu's departure from the White House, they were able to bring about a softened U.S. posture with regard to UNCED and international environmental equity generally. However, they were unable to have enough influence to overcome most of Sununu's influence. At least one American delegate with a reputation for being sympathetic to the South's equity demands was eventually shunted by Sununu "appointees" to sideline talks, specifically in order to minimize her effect on the central UNCED negotiations dealing with finance and technology transfer.[71]

Throughout the UNCED process, other developed country governments—not to mention the developing country governments—were frustrated with the U.S. refusal to take a leadership role on global environmental issues generally and on the demands of the developing countries in particular. To be sure, many other industrialized countries viewed many of the more strident demands for international equity with skepticism, but they also realized that some of the demands were reasonable, given the context of global environmental commons issues, and in some cases they (e.g., the Nordic states, in particular) interpreted many demands for international equity as justified.[72] At the least, other developed country governments were frustrated that the United States was unwilling to discuss such issues at first and was generally uncooperative in efforts to engage the developing countries.

The bewilderment of U.S. allies was manifested right through the negotiations on climate change, for example. Foreign participants in the April 1990 White House meeting on climate change were frustrated by the lack of results and vented those frustrations to the international press. Hans Alder, Environmental Minister from the Netherlands, said that delaying action to address climate change would only increase the long-term costs and that "no one had the right to conduct experiments on a global scale and transfer risks to the vulnerable countries and future generations."[73] According to Hatch, the conference hardened the position of the Europeans and gave domestic environmental groups and Congress "a golden opportunity to highlight the shortcomings (from their perspective) of the administration's position on global warming."[74]

Throughout the negotiations on the Climate Change Convention the United States faced off against other OECD countries pushing for carbon dioxide emissions targets and timetables. For example, at the March 1991 Geneva prepcom, delegates from other industrialized countries publicly isolated the United States on the climate change and forest issues, with especially strong criticism from British Environment Minister Michael Haseltine for U.S. resistance to stricter curbs on carbon dioxide emissions, and from EC director general for the environment Laurens Binkhorst, for the U.S. government's failure to recognize inordinate American emissions of carbon dioxide.[75] In the early phases of negotiations on the Earth Summit agenda, there were, according to Brenton, "unavailing continental European efforts to push or embarrass the U.S. into changing its position. Occasionally these pressures produced tactical concessions—mostly intended to placate U.S. public opinion—but on the main point, opposition to carbon dioxide emissions targets, the U.S. was immovable."[76] Indeed, Bush linked his participation in the Earth Summit with progress made in the INC toward satisfying his concerns on this issue.[77] The widespread criticism from outside the United States was manifested in the decision of Carlo Ripa di Meana, the European Community's environment commissioner, to boycott the Earth Summit because the United States had vetoed numerical goals for greenhouse gas emissions reductions.[78]

The U.S. delegation arrived at the Earth Summit with their government viewed as a "villain," "party pooper," and "Uncle Grubby." Even U.S. allies were suspicious of the U.S. Government and thought it was merely "playing games."[79] The U.S. delegates experienced anti-American sentiment reminiscent of the Vietnam War. The United States was strongly criticized in the world press for refusing to sign the Biodiversity Convention. The official newspaper of the Earth Summit, *The Earth Summit Times*, declared that the United States was on a "slide into isolation at the Earth Summit."[80] A Brazilian congressman participating in the summit said that U.S. intransigence was "recreating the polarized atmosphere of the 1960s: all civil society and the press against the U.S."[81] "For the Bush Administration," according to Scott Hajost (a U.S. delegate to the UNCED negotiations), "Rio was an unmitigated public relations, foreign policy and political disaster that was not forgotten in the '92 elections."[82] Ironically, the negative posture of the U.S. government may have contributed to the political solidarity of many Southern and Northern (especially EC) governments more supportive of UNCED's equity goals.[83]

The actors who tried to influence the Bush government's UNCED policies operated in the context of the emerging international consensus on international environmental equity. When the Bush administration took a strong position against UNCED considerations of international equity, it was acting in opposition to the emerging consensus, which was that many of the developing countries' demands for international equity were reasonable, and that at least in the environmental issue area it was in everyone's interest for the North to be more forthcoming with regard to those reasonable demands. This emerging consensus was manifested in international environmental institutions, notably the UNCED Secretariat and the UN generally.

International institutions and regimes, and the organizations associated with them, contributed to more prominent considerations of international equity in the UNCED deliberations. Levy, Keohane, and Haas describe how institutions contribute to bargaining environments like the UNCED deliberations: International institutions facilitate linkage of issues; create, collect, and disseminate scientific knowledge; create opportunities to magnify domestic public pressure; provide bargaining forums that reduce transaction costs and create iterated decision making processes; conduct monitoring of environmental quality, national environmental policies and performance; increase national and international accountability; and boost bureaucratic power of domestic allies.[84] These avenues of influence were germane in the U.S. case.

The institutions and organizations most noteworthy for promoting equity considerations as a way of fostering more effective international environmental cooperation (or as an objective for other reasons) have been the United Nations system generally and the UN Environment and Development Programs in particular, the nascent climate change and other environmental institutions formed in the years prior to the Earth Summit, and the foreign aid regime that formed in the years after the Second World War. International organizations initially pushed the environment-development agenda. The United Nations Environment Program, for example, initiated negotiations on climate change and biological diversity, much as it had helped stimulate the international deliberations on stratospheric ozone depletion. Gunnar Sjostedt and colleagues point out that, unlike the GATT negotiations, in the UNCED negotiations leadership of the great powers (e.g., the United States and the EC) "was largely lacking, particularly in the early stages of the negotiation process when issues were framed and the agenda set. In this phase of the UNCED process, negotiations were almost entirely dependent on the input supplied by the UNCED secretariat and the Bureau, including the chairman of the prepcom as well as the chairmen of the working groups."[85] Thus, UN agencies were able to frame the debate in which the Bush administration found itself. Because the White House took interest in most of the UNCED deliberations only late in the process, it was already waging an uphill battle against the emerging consensus that equity considerations had to be part of effective efforts to address environmental change.

While the organizational manifestations of international institutions—in this case primarily UN organs—did not have a profound direct influence on the Bush administration's policy on international environmental equity, they did have an important indirect impact on that policy by ushering the notion of sustainable development and international equity into international environmental discourse. They facilitated the raising of questions about equity in international environmental politics. The UN and the UNCED secretariats—bolstered by the developing countries—kept the deliberations focused on the connections between environmental protection on the one hand, and economic development and poverty eradication in the South on the other. Maurice Strong, Secretary General of UNCED, traveled around the world in what one observer called an "evangelical mission to convert the world to the cause of the Earth Summit and sustainable development."[86] Although Strong's effect on the Bush administration appears to have been limited, the U.S. government was operating in a bargaining environment already permeated by notions of international environmental equity. While the United States usually found such ideas unpalatable, it could not avoid engaging in discussions about them on a recurring basis.

The Politics of Climate Change and Biodiversity During the Bush Administration

During the UNCED and climate INC negotiations, the United States, alone among the OECD countries, refused to accept binding targets and timetables to reduce carbon dioxide emissions to 1990 levels by 2000. According to Scott Hajost, this was the U.S. position despite the Bush administration's own analysis showing that such a commitment was achievable. The U.S. government had an *ideological* objection to any form of international targets and timetables, which it had also strongly opposed in the UN Convention on the Law of the Sea.[87] But even the United States was unwilling to go as far as some countries interested in preventing action on climate change altogether. During UNCED negotiations on Chapter Nine of Agenda 21 (dealing with protection of the atmosphere), after recommendations on energy efficiency and least-cost pricing were weakened, the United States, "in the final hours of the Main Committee, finally said 'enough is enough' to Saudi efforts to weaken further the text by deleting references to efficiency and renewables. Arguably, only the United States was in the position to make this point and allow Tommy Koh, Chairman of the PrepCom and Main Committee Chair, to gavel the Saudis down."[88] Strangely, considering its opposition during the climate talks, the Bush administration sent the treaty with great enthusiasm to the Senate, and in October 1992, the United States became the first developed country (and third overall) to ratify it.[89]

According to Michael Hatch, the U.S. position on the burdens (targets and timetables) and benefits (for developing countries, in the form of new funds and concessional technology) of agreements on climate change were results of American domestic politics, international negotiations, and the nature of the climate change issue itself (its complexity, the many interests it affects, and the requirement that all major countries cooperate in addressing it).[90] A disparate and broad set of interests was activated by climate change concerns. Those interests had access to the decision making process as a result of the pluralism, of U.S. politics. The consequence was that a firm consensus on climate change in the U.S. government could not be reached. Domestically, congressional committees and members of congress, bureaucratic agencies and departments, and influential individuals had competing interests and notions of how the United States should address the climate change question. These entities and individuals were in turn influenced by foreign governments and interests. At times international influences strengthened domestic groups competing in the domestic policy process. However, according to Hatch, while international forces were important and did affect U.S. policy on climate change, domestic politics severely circumscribed the ability of the U.S. government to obligate itself to firm targets and timetables for reduction of carbon dioxide and other greenhouse gases. Yet, by the end of the Reagan administration, climate change had become a front-burner issue. Hatch summarizes:

> In the course of the 1988 presidential campaign, George Bush felt compelled to utter his now famous statement that "those who think we are powerless to do anything about the 'greenhouse effect' are forgetting about the 'White House Effect'"; members in Congress from both sides of the aisle were actively engaged in the issue and equally intent on engaging the

executive branch in analyses of the greenhouse effect and what could be done to reduce it; several departments and agencies of the executive branch brought their own particular sets of interests and institutional perspectives to the global warming questions; and individuals as well as groups from the scientific and environmental communities were pressing their concerns on both branches of government.[91]

Both the Department of State and the EPA had become important participants in the domestic politics of climate change. But Secretary of State James Baker would eventually recuse himself from the issue (perhaps due to the potential conflict of interest from his oil and gas holdings[92]), reducing the State Department's influence. The EPA administrator, William Reilly, would find it extremely difficult to influence Sununu and his allies opposed to U.S. action on climate change. A split in the administration formed and was evidenced at the Noordwijk climate change conference. The Netherlands, Germany, France, Canada, Sweden and Norway called for agreement on binding commitments to stabilize carbon dioxide emissions by 2000 and to reduce them 20 percent by 2005. The EPA and State Department supported a stabilization target, but Sununu and representatives of the Department of Energy and the Office of Management and Budget (OMB) were opposed. Concerned about the potential costs of climate change for the United States, skeptical of the science, and concerned that international events were pushing the United States toward action, Sununu and Richard Darman, director of OMB, increased their efforts to control the policy process. At the Noordwijk conference the United States (with Japan and the USSR) in effect opposed quantitative targets.[93] This was indicative of successful White House efforts to expand its influence beyond domestic politics to shaping international negotiations as well.[94] The U.S. policy through 1991 was, in short, opposition to any binding targets or timetables for the emission of greenhouse gases[95] (with exception of agreements made to reduce emissions of CFCs in the context of the Montreal Protocol process), opposition to assuming responsibility for past emissions, and opposition to binding obligations to provide new funds for developing country efforts to limit greenhouse gas emissions.

However, beginning in 1992 the Bush administration's position began to change toward some recognition of international environmental equity considerations. According to Hecht and Tirpak, the United States would not accept binding targets and timetables, but it "made a major concession on financing at the February INC meeting. At the urging of senior officials of the Treasury, EPA and OMB, the President made a decision to change U.S. policy with regard to financial assistance to developing countries. The United States for the first time acknowledged the need to provide financial resources to developing countries to permit their full participation in the Convention."[96] The United States reversed its well established policies by pledging to contribute $75 million to help developing countries stabilize their greenhouse gas emissions, with two-thirds of this funneled through the GEF. Furthermore, the U.S. delegate announced a set of measures to reduce U.S. emissions of greenhouse gases by amounts that would "compare favorably with those of other developed countries."[97]

Bush administration officials told the media that further movement on the U.S. position before the Earth Summit was likely.[98] Indeed, the administration's April 1992 report describing "U.S. Views on Climate Change" seemed to provide a basis for fundamental revisions to the negotiating position on climate change obligations. The report (produced by a working group with members from four agencies) indicated that the United States could come close to the Europeans' emissions targets at little economic cost by implementing then current energy policies along with additional voluntary measures. However, in the final draft of the Climate Change Convention the U.S. government remained opposed to binding commitments and specific targets and timetables—despite agreeing to compromise its most hard-line position after several British diplomats went to Washington to negotiate treaty wording acceptable to both the United States and the Europeans.[99] "Nonetheless," according to Hatch, "the treaty represented significant change when measured against positions staked out earlier by the Bush administration [which] now came to accept the basic science of global warming and its potential dangers. Accordingly, policy focused less on whether climate change was a problem, more on how best to address it. Finally, the treaty agreed to by the U.S. seemed to provide certain criteria, albeit nonbinding, against which national action can and should be measured."[100]

What explains these changes? According to Hatch, they were the result of three factors: (1) changes in White House personnel, most notably the departure of Sununu; (2) the presidential election campaign; and (3) international pressures, especially from other developed countries.[101] Sununu had taken control of the climate change debate in the Bush administration. After he left government a more disparate set of views were discussed by administration officials, including pro-environmental perspectives that were expressly suppressed by Sununu. Environmental interest groups were invited by the new chief of staff to join in high-level discussions on climate change. According to a senior U.S. diplomat involved in the UNCED negotiations, it is unlikely that the U.S. government would have signed the Climate Change Convention if Sununu had stayed on.[102]

In addition to the so-called "Sununu Factor," changes in the U.S. negotiating position away from the pre-1992 hard-line policy can be attributed to pressures from other developed countries. The "mellowing" of the U.S. position vis-a-vis the developing countries' demands had much to do with contacts the president and other high level officials in the Bush administration were having with individuals in other developed countries.[103] For example, at the G-7 summit at London in 1991, Bush was pushed by other G-7 leaders to attend the Earth Summit. Condemning the U.S. failure to lead on climate change, British Prime Minister John Major said: "The United States accounts for 23 percent [of global carbon dioxide emissions]. The world looks to them for decisive leadership on this issue as on others."[104] On funding, the EC and Scandinavian countries insisted that new and additional funds and technology transfers were essential to an effective convention.[105] Shortly before the Rio conference, the British negotiators went to Washington to negotiate a compromise that codified the more flexible position and made it possible for Bush to attend the summit. According to two EPA officials, for many European governments emissions targets and timetables were a symbolic challenge to U.S. leadership as the EC tried to become the center of influence on environmental issues.[106]

Additionally, according to Hatch:

> Bureaucratic actors may have found the actions and views of other countries useful in bolstering their position. More telling was the sensitivity of administration officials to the efforts of environmental groups attempting to portray the United States as standing alone among the developed countries in the talks. . . .these same officials often made reference to countries that publicly advocated firm targets and timetables for carbon dioxide emissions. . . . In other words, the administration took seriously attempts to shape public perceptions of the U.S. position in the international negotiations through reference to the other major countries.[107]

The result was a more flexible and positive U.S. position, one that viewed U.S. measures to reduce carbon dioxide emissions as possible at reasonable cost, and one that contributed to subsequent agreement by the United States to wording found in the FCCC.[108] Hence, U.S. policy shifted during the course of negotiations from outright opposition to action to accepting "no regrets" and "comprehensive approach" policies that were ultimately accepted by the EC countries and incorporated into the final agreement.[109] Bush pledged that the United States would quickly ratify the treaty and develop a national action plan by January 1993.

Ideological and economic concerns in the White House and presidential campaign politics explain why the departure of Sununu and international pressures were not strong enough to move the Bush administration closer to the position of the Europeans on targets and timetables (i.e., binding commitments as opposed to a flexible pledge).[110] Other administration officials, especially in the OMB (most notably director Richard Darman) and the Council of Economic Advisers (especially Chairman Michael Boskin), viewed changing U.S. dependence on fossil fuels as too economically costly, and they rejected government intervention in the economy that might be necessary to address climate change.[111] They were also cognizant of the political power of the pro-business Global Climate Coalition, Citizens for a Sound Economy, and other organizations opposed to stabilization of carbon dioxide and other greenhouse gas emissions. On the other side were environmental interest groups that would measure Bush's performance as the "environmental president" by his actions at the Earth Summit. Democratic presidential candidates made climate change a prominent issue in their campaigns; all the Democratic candidates supported carbon dioxide emissions targets. Bill Clinton stated his support for a climate change treaty in which the United States would agree to stabilize its carbon dioxide emissions at 1990 levels by 2000, and he added that the United States should make further reductions.[112] Rather than move toward the Democratic and environmentalist position, however, Bush chose to hold his ground to garner support from the most conservative Republicans and Reagan Democrats.[113]

The United States, in signing the FCCC, "signaled an acceptance of positions long resisted by the Bush administration. Nonetheless, the domestic political process, combined with the nature of the global warming issue itself, limited the amount of change possible."[114] The pluralist nature of American politics, characterized by the constitutional separation of powers, a fragmented congressional committee system, and numerous access points for disparate interest groups, was magnified by the complexity and widespread effects (both environmental and economic) of the climate change problem.[115] As a result, the Bush administration's policy on climate change reflected a modest shift toward acceptance of international equity—more than it had planned to do, but less than what the Clinton administration would eventually do.

Similar processes were at work in the development of U.S. foreign policy on biodiversity. During the Bush administration, both the public at large and the U.S. government were increasingly interested in protecting threatened species in the United States and abroad. The U.S. government slowly come to recognize that extinctions of species—the loss of biodiversity—had the potential to harm Americans and the U.S. economy in unpredictable ways. There was growing concern that loss of biological diversity could reduce genetic diversity, damage the earth's atmosphere, and adversely affect supplies of fresh water, fish, and forests.

A variety of disparate actors with disparate interests were concerned about species loss: the public, environmentalists, foresters, agronomists, biotechnologists, and pharmaceutical and other industries.[116] A shrinking genetic pool could mean the loss of material useful for agriculture, medicines, and a variety of new biotechnologies modeled from natural species. By 1992 about 25 percent of the prescription drugs used in the United States had active ingredients derived from plants, many from developing regions of the world. The dollar value of pharmaceuticals derived from genetic materials originating in tropical forests was estimated to be $22 billion annually at the start of the 1990s.[117] Thus the loss of species could have both health consequences for U.S. citizens and economic consequences for U.S. industry.[118] As Brenton pointed out, "While the vast majority of the species being destroyed are not of the politically emotive sort which generated earlier conservation efforts, being insects, fungi and plants rather than elephants, whales and birds, they do have a real economic and ecological importance."[119]

Despite these concerns, the Bush administration developed a strong opposition to the Biodiversity Convention, and days before the Earth Summit it announced that the United States would not sign—despite the fact that the United States had first introduced the idea of a treaty to protect species. The Bush administration was responding to pressure from industry groups, including the Industrial Biotechnology Association, the Association of Biotechnology Companies, the Industrial Pharmaceutical Association, the Pharmaceutical Manufacturer's Association, and the American Intellectual Property Law Association, which were opposed to wording in the Convention that called for technology transfer to (primarily developing) countries that were sources of genetic resources. According to Frye, Lisa Raines, Vice-President for Industrial Relations of the Industrial Biotechnology Association, contacted the U.S.'s chief negotiator on the Convention and requested that all references to intellectual property rights be removed from the treaty. Patent lawyers lobbied the Bush administration, pushing it to make changes in the intellectual property provisions of the Convention. According to Frye, the biodiversity treaty was compared to the Law of the Sea Treaty in which the United Nations would have had the authority to transfer deep seabed mining technology to developing countries. The Reagan administration would not sign the treaty because it would require American companies to transfer technology in a "forced sale."[120] So too with the Bush administration and the Convention on Biological Diversity.

The U.S. government's main objections to the Convention were threefold: (1) the treaty's provisions dealing with biotechnology and the associated demands by developing countries for new funding and access to biotechnology on preferential terms; (2) the likelihood that Convention would lead to developing countries patenting genetic material, thereby requiring American biotechnology firms to pay royalties on products made from that material; and (3) financial provisions of the Convention that would permit the developing countries to control decisions regarding the amount of funds required by the Convention, and the related issue that developed countries would be legally obligated to provide whatever funds the developing countries determined were necessary under the Convention. The Bush administration thought that the Convention as written would restrict the application and industrialization of biotechnologies, jeopardize protections for intellectual property rights, and reduce royalties for pharmaceutical companies.[121] Frye described the opinion among conservatives in the United States regarding the Biodiversity Convention: "[M]any of the technology transfer types of programs sought by the developing nations would be sort of a 'double whammy' for Americans: after having spent a substantial portion of our income to install and improve pollution controls on U.S. industry, which also has resulted in the loss of many jobs in heavy industry to other countries, Americans are now asked to fund pollution control measures by the countries with which we compete, and to transfer to them on favorable terms the high technology products that we are still able to produce competitively."[122]

These concerns prevented the Bush administration from signing the treaty. One reason it was able to get away with this was the low profile of the negotiations. Compared to the effort put into the climate change negotiations, interested parties within the administration devoted relatively little time, money, and personnel to the process. The EPA (especially its administrator, William Reilly) and the Department of State (especially the leader of the U.S. negotiating team, Assistant Secretary of State E.U. Curtis Bohlen) supported the biodiversity treaty. Alternatively, White House staff considered the Biodiversity Convention as "just an international endangered species act."[123] Vice President Dan Quayle signed a memorandum attacking the draft biodiversity treaty, saying that it would "hamper the U.S. biotechnology industry, greatly expand the reach of the Endangered Species Act and force the United States to enact a host of other regulations."[124] Differences within the administration became public during the Earth Summit when a confidential June 3 memorandum from Reilly to Clayton Yeutter at the White House was leaked to the press. Reilly wanted the United States to agree to minor modifications in the Biodiversity Convention and then sign it. However, a staff member in Vice President Dan Quayle's highly conservative, anti-regulatory, and pro-business Council on Competitiveness (apparently intent on embarrassing Reilly, an environmentalist who was continually at odds with right-wing conservatives in the Bush administration[125]) leaked the memo. The White House refused Reilly's request.

The Vice President's Council on Competitiveness was able to take control of the biodiversity agenda in the Bush administration, in large part because the negotiations garnered little public scrutiny or concern. The Council argued that the treaty would harm the U.S. economy by forcing American businesses to follow new international regulations and requirements, and to share their ideas and technologies with Southern competitors.[126] The United States was suffering from an economic recession in 1992, thus making these types of arguments especially germane. The Bush administration, in trying to balance economic interests of the United States, heard the opponents of the convention much more loudly than it heard those who were advocating U.S. participation in a strong biodiversity treaty.

Some biotechnology companies, such as Merck and Genentech, interpreted the convention as favorable to their long-term corporate interests. It would, they hoped, help create a cooperative climate whereby deals could be reached with developing countries to ensure access to biological and genetic resources. In 1991 Merck signed a contract with Costa Rica's National Biodiversity Institute, which agreed to transfer insect, plant and animal materials to Merck for two years in exchange for a small percentage of the royalties from any pharmaceuticals that might be developed from those materials, along with a lump sum of one million dollars. These companies would later ally themselves with environmental organizations to pressure the Clinton administration into signing the Biodiversity Convention.

Other developed country governments did not share the Bush administration's strong objections to transfer of technology and access to biotechnology, and therefore signed the Convention (the British signed subject to an interpretive declaration limiting the power of the Conference of the Parties to determine funding priorities).[127] In many respects it boiled down to two incompatible perspectives taken by developed countries: The United States was clearly interested in protecting biological diversity (if for no other reason than to protect genetic sources of new biotechnologies), but was unwilling to sacrifice its freedom of action and the patent rights of its corporations. Other developed countries saw the Biodiversity Convention and the concessions made to developing countries that possess much of the world's genetic resources as a price worth paying to ensure the protection of those resources.

Political Forces Acting on the Clinton Administration

The actors that influenced the Bush administration's policies with regard to international environmental equity also influenced the Clinton administration's policies. Their policies developed in the similar contexts: the pluralist nature of American politics and policy making, whereby numerous disparate actors were able to access the policy process and affect the actions of politicians concerned about reelection (not to mention the actors' access to the federal bureaucracy), and the emerging international consensus on international equity in the environmental field. In addition, the Clinton administration was affected by what happened to the Bush administration. Clinton did not want to be seen as a politician consorting with businesses bent on destroying the environment; quite the contrary: he wanted to be seen as a moderate environmentalist who would work with business to protect the environment. The actors and forces that helped transform the Bush policy from one of absolute opposition to international equity considerations into a policy that ranged from non-obstructionist to mildly supportive in specific instances, also helped transform the Clinton policy from cautious support into relatively strong acceptance of international environmental equity as a principle of U.S. foreign policy (albeit with only modest follow up). Hence, many of the explanations for the Clinton administration's policies are the same as those for the Bush administration.

However, the Clinton administration was influenced by the various actors differently than was the Bush administration. Whereas some industries pushed Bush to oppose international environmental equity in the context of UNCED, other (and sometimes the same) industries put pressure on a more sympathetic Clinton administration to support equity considerations. Whereas powerful individuals in the Bush administration (e.g., John Sununu) were successful in preventing the U.S. government from accepting more actively international equity as an objective of global environmental policy, in the case of the Clinton administration powerful individuals (e.g., Al Gore) had the opposite impact. Whereas scientists skeptical about climate change got the ear of the Bush White House—and environmentalists did not—scientists (drawing on the resources of the IPCC) and concerned environmentalists were heard by the Clinton administration (indeed, many of them joined it).

Much like the Bush administration, the Clinton administration was pulled in opposing directions by its concerns for reelection. On the one hand, it wanted to appeal to environmentalists—organizations and citizens (more specifically, voters) sympathetic to the acceptance of international environmental equity by the U.S. government. Alternatively, it wanted to avoid upsetting business interests that remained opposed to many equity-related provisions of international environmental agreements, especially the Climate Change Convention. However, unlike the Bush administration, the Clinton administration was not campaigning during an economic recession, and new coalitions of business organizations had arisen to counter the influence of those groups opposed to the Climate Change Convention. Many business groups once opposed to the biodiversity treaty also changed their position and were now pushing for its ratification. Nevertheless, the Clinton administration chose not to undertake radical changes in U.S. global environmental policy before the 1996 elections, although its *proposals* were quite radical, most notably its endorsement of a legally binding international agreement to reduce greenhouse gas emissions.

The Bush administration faced a Congress largely sympathetic to UNCED and the equity provisions of its agreements and treaties; in short, it faced a Congress opposed to the thrust of its preferred policies. With the Republican takeover of Congress in 1994, the Clinton administration also faced strong opposition to its policies from Congress, but this time the Congress was generally opposed to the UNCED agreements and especially provisions that would require the United States to increase foreign aid, give more power to the developing countries (especially with regard to spending money donated by the United States), transfer technology on concessional terms, or take full responsibility for its greenhouse gas emissions. In short, through an ironic reversal of what occurred during the Bush administration, Congress was one of the main reasons why the Clinton administration was not able to implement its preferred policies—including a stronger endorsement of international equity as an objective of global environmental policy—and why it was unable to create more new sustainable development initiatives that required substantial new funding.

The Clinton administration received mixed signals from the public at large on international environmental issues. On the one hand, much of the public wanted the administration to do more to protect the environment, and this was reflected in the administration's efforts to prevent the

Republican-controlled Congress from reversing domestic environmental regulations dating from the 1970s. Many Americans were also generally aware of and concerned about the dangers posed by climate change, tropical deforestation, and ozone depletion (although they did not necessarily understand *why* these things were potential threats).[128] On the other hand, the public was generally opposed to increased foreign aid, being grossly uninformed about the size of the foreign aid budget relative to the overall federal budget. One of the mantras of the mid-1990s was deficit reduction. A majority of Americans believed that foreign aid spending was several times higher than actual levels, and many Americans mistakenly thought that by limiting foreign aid the federal budget deficit could be substantially reduced.[129] Consistent with this public ambivalence, the Clinton administration displayed a willingness to join international efforts to protect the environment, and it accepted international equity as an important means to achieving environmental protection, but rather than press for substantial new funds from a Congress opposed to increased foreign aid, it chose to divert resources from existing programs toward new sustainable development and poverty eradication programs for the poorest countries. It also worked with other governments, international financial institutions (notably, the World Bank), nongovernmental organizations, and industry to ensure that official development assistance and private investment alike were more attuned to the equity provisions of the international environmental agreements devised under the rubric of UNCED.

Much as during the Bush administration, the Clinton administration was subject to intense pressure from commercial interests. However, the coalition of forces in the latter case was much different and much more conducive to the U.S. government's acceptance of international environmental equity and the international agreements in which it was codified. In the case of climate change, while the same forces remained arrayed against action on U.S. consumption and binding agreements to limit emissions of greenhouse gases, the scientific support for opponents' positions was much weakened. Just as important, if not more so, other business interests lobbied the Clinton administration to endorse agreements that reflected the emerging international consensus on international environmental equity. The insurance industry joined forces with environmentalists and the environmental and energy technology industries to persuade the Clinton administration to *support* a binding agreement to address climate change. Similarly, environmentalists and the growing biotechnology industry lobbied for signing and ratification of the Biodiversity Convention (having interpreted it anew and seen it as promoting U.S. interests).

Pressure from nongovernmental organizations was intense, but in the case of the Clinton administration NGOs were trying to influence a much more sympathetic White House. Indeed, many of the people working on global environmental issues in the Clinton administration came from environmental NGOs. Such groups were brought into the policy-making process, and they were seen by the Clinton administration as part of cost-effective solutions to environmental and development problems. The old-line U.S. conservation groups—the Sierra Club, the National Audobon Society, the National Wildlife Federation—had expanded their mandates to include global environmental issues *and* economic development goals.[130] Moreover, the constituencies of these types of NGOs were much more important to Clinton than to Bush. The post-Rio "democratic" restructuring of the GEF—and changes made in the larger World Bank environmental funding programs—reflected the concerns of American NGOs who lobbied the Clinton administration, which in turn pressed other donors to agree to changes in how the GEF operated.[131] Nongovernmental organizations had long criticized the World Bank for its closed decision making process, and for its funding of huge infrastructure projects that often did more harm than good and too rarely helped to alleviate poverty and environmental damage at the local level.[132] Reflecting the Clinton administration strategy of trying to engage the whole range of interested actors in its national and global environmental policies, the new President's Commission on Sustainable Development included representatives from not only government, but also business and environmental groups.

Scientists calling for international regulations to protect the environment were not viewed favorably by the Bush administration. The 1990 first assessment report of the Intergovernmental Panel on Climate Change, which did not even state that climate change was being caused by human activities, nevertheless met with great skepticism from the Bush administration. The Clinton administration was decidedly less skeptical, acknowledging that the evidence was building with regard to human impacts on climate. Then came the second report of the IPCC, issued at the end of 1995, indicating that there was a "discernible human influence" on climate. The report said that the earth's global mean temperature would likely rise between 1 and 3.5 degrees Celsius by the year 2100, and that temperatures could continue rising thereafter even if emissions of greenhouse gases were stabilized in the interim.[133] This increased the pressure on the Clinton administration to acknowledge human, and particularly Americans', adverse impact on global climate. As a result, there was a new consensus on climate change within the government that did not exist during the Bush administration.

During the Clinton administration, certain well-placed individuals played very important roles in shaping U.S. policy on international environmental equity. Vice President Al Gore in particular imposed himself on most aspects of U.S. global environmental policy, admitting the U.S.'s over-consumption relative to the rest of the world, favoring new aid to the South for sustainable development programs, and supporting the Biodiversity Convention and targets and timetables for greenhouse gas emissions reductions. What is more, Gore participated in the appointment of like-minded individuals throughout the environmental foreign policy bureaucracy. Former Senator Timothy Wirth, one of Gore's chief allies during the UNCED negotiations and the Earth Summit, was appointed to a new State Department post created to deal with global issues, including environmental change and sustainable development. Rafe Pomerance, a former environmental activist, and formerly president of Friends of the Earth and a senior associate of the World Resources Institute, was appointed to head the first Clinton administration's climate change delegations as deputy assistant secretary of state for environment and development. Pomerance had been pushing for a decade before the Earth Summit to persuade the international community that climate change posed a serious threat.[134] Many other individuals with activist environment and development backgrounds became part of the Clinton administration. The Clinton administration was also instrumental in the 1993 appointment of James Gustave Speth, a lifelong American environmentalist, former president of the World Resources Institute, and chairman of President Jimmy Carter's Council on Environmental Quality, as head of the United Nations Development Program.[135] Speth promptly made sustainable development in the world's poorest countries a guiding principle of UNDP.

The Clinton administration was also influenced by foreign actors. It sometimes listened to the arguments of the developing countries for new funding, technology, reduced consumption in the United States, and related matters, and it was sympathetic to many of those arguments (up to a point) because it knew that the developing countries' cooperation would be required to effectively address environmental issues important to the administration. Even so, the Clinton administration was unwilling to introduce legislation that would require U.S. corporations to hand over environmental technologies to other countries, although it did work to help fund such transfers on mutually agreed terms and to improve access to the bulk of technologies (which were often already in the public domain).

The Clinton administration felt pressure from other developed countries on the issue of development assistance. Like the United States, they were reducing their aid budgets, but their per capita giving remained higher—in several cases very much higher—than that of the United States. They urged the United States to join them in providing more aid to the world's poor. Reflecting the sentiments of the Western European countries, in a 1996 speech before a joint session of the U.S. Congress, French President Jacques Chirac called on the United States to give more money to promote economic development in the world's poorest countries. He pointed out that the European Union, with an economy about the same size as the United States, provided about three times as much official development assistance each year.[136] Due to congressional opposition and a public skeptical of foreign aid, however, new overall funding was impossible, although the administration did divert more aid to sustainable development and poverty eradication programs. Thus foreign pressure may have been somewhat effective in influencing the Clinton administration, but much less so in the case of the U.S. government overall. In any case, much of what the Clinton administration was hearing from abroad it was already hearing from Gore and others within the administration.

The views expressed by foreign governments frequently reflected the emerging international consensus on international equity in the global environmental field. This consensus was manifested in international environmental institutions and organizations, which had an indirect impact on Clinton administration policy. The UNCED follow-on discussions regarding climate change, biodiversity, desertification, oceans and the like were held under the auspices of these institutions. American diplomats and advisors were in ongoing contact with secretariats of various UN agencies and international environmental institutions, especially those created by UNCED (the Global Environment Facility, the Commission on Sustainable Development, the Climate Change Convention, and the Biodiversity Convention). Many of the follow-on discussions occurred in meetings of the Commission on Sustainable Development, which was established to ensure, at least in part, that equity issues would be incorporated into all United Nations-sponsored efforts to promote development and environmental protection, especially in the poor countries. The Clinton administration took a much more positive approach to the UN and international environmental institutions and organizations than did the Bush administration. It often preferred multilateral approaches to foreign policy problems, especially those concerning environment and development.

The Politics of Climate Change and Biodiversity During the Clinton Administration

The Clinton administration took a more sympathetic position on international equity as it relates to climate change issues and the Framework Convention on Climate Change.[137] In the 1990 presidential campaign, Governor Clinton and Senator Gore said that it was in America's interest to use less energy and to stabilize U.S. emissions of greenhouse gases at 1990 levels by 2000. They also said that if elected their administration would consider greater reductions of such emissions in the future. After taking office, Clinton and Gore continued to support efforts to address climate change, at least rhetorically. Vice-President Gore told the House Foreign Relations Committee that protocols to the FCCC should be negotiated quickly. At the first substantive meeting of the Commission on Sustainable Development in 1993, he said that President Clinton recognized the need to reduce emissions of greenhouse gases and that Clinton—in what Gore called "a major change for my country"—had committed the U.S. government to reducing emissions to 1990 levels by 2000.[138] Slowly, the United States was agreeing that it ought to take on a greater share of the burden of addressing a problem that was in large measure of its own making. However, the Clinton administration was initially opposed to *legally binding* targets and timetables to meet the Climate Change Convention emissions goals.

During the Bush administration the science of climate change was highly uncertain, making the calls by White House chief of staff Sununu and industry officials to do more research before taking any action to reduce greenhouse gas emissions seem almost reasonable (short of "no regrets" actions, to which the administration eventually agreed, albeit reluctantly). After Bush left office, however, atmospheric scientists improved their data and climate models, and some industries joined the majority of scientists in calling for action to reduce greenhouse gas, especially carbon dioxide, emissions. In late 1995, 2,500 scientists of the UN's Intergovernmental Panel on Climate Change reached consensus in their second assessment report that climate change is indeed a threat, that its consequences are uncertain but will be harmful overall, that it is caused in part by human activities, especially the burning of fossil fuels, and that reductions of greenhouse gases on the order of 50 to 70 percent would be required just to stabilize atmospheric concentrations.[139] This strengthened the hand of those urging the United States to promote international equity as a necessary and effective way of persuading developing countries to join the emerging climate regime. Thus, whereas during the Bush period the administration, some scientists, and the coal, oil and automobile industries (in the form of the Global Climate Coalition) were arrayed against environmentalists and other scientists, in the Clinton period the administration was increasingly allied with most atmospheric scientists, environmentalists, environmental technology businesses, and the assertive (and wealthy) insurance industry.

The insurance industry, fearing huge financial liabilities from climate change, joined with those supporting the FCCC.[140] The industry acted as a new counterweight to the fossil fuel and auto industries. (Globally, insurers paid out tens of billions of dollars for damage incurred from storms that were increasingly frequent and severe in the late 1980s, and which insurers believed were caused by climate change. Some insurers were forced out of business, and some predicted that climate change could bankrupt the insurance industry in the United States and worldwide. It was estimated that a single hurricane in the United States could cost the global insurance industry $50 billion.[141]) Insurers had talks with Vice President Gore, placed advertisements in major media, and even joined with the activist environmental organization Greenpeace to produce a book on climate change, in addition to doing more traditional things like lobbying for stricter building codes and generally pushing the government to take action to reduce greenhouse gas emissions and thereby limit climate change.[142]

In the post-Rio deliberations of the FCCC's conferences of the parties, U.S. industries hoping to promote alternatives to fossil fuels lobbied meetings and "sharply reduced the influence of pro-fossil fuel industries, which had previously monopolized industry views on the issue."[143] There were predictions that the market for new environmental technologies could reach $400 billion annually,[144] and U.S. corporations wanted the government to support international agreements that would provide incentives for the purchase of those products, such as climate change mitigation measures that would bolster demand for energy efficient products, manufacturing processes, and electricity plants. The Business Council for a Sustainable Energy Future, a coalition of European and U.S. business leaders representing clean energy industries, declared their full support for the IPCC second assessment report and for an international agreement to cut greenhouse gas emissions. Micheal Marvin, executive director of the group's U.S. branch, which consisted of over 200 businesses, said that action to reduce greenhouse gas emissions was required "because the cost of no action could be devastating" (it is unclear whether he was referring to his industry, the United States, or the world).[145]

Nongovernmental organizations also pressed the Clinton

administration to be more forthcoming with regard to U.S. consumption and greenhouse gas emissions reductions. For example, at the July 1996 second meeting of parties to the FCCC, U.S.-based environmental organizations numbered in the dozens. They included Greenpeace, the World Wildlife Fund, the Environmental Defense Fund, and the Climate Action Network (a large coalition of environment and development NGOs), and many others. Pressing the Clinton administration from a somewhat different angle, the National Council of Churches, the United States' largest ecumenical organization with 52 million members, met with Clinton administration officials and members of Congress in 1996 to persuade the U.S. government to reduce domestic emissions of greenhouse gases to 1990 levels by 2000 and to support a binding international agreement that would require developed countries to reduce their greenhouse gas emissions after 2000. General Secretary of the council, Joan Brown Campbell, said the risk of human-caused climate change involved "profound issues of global justice. The potential impacts of climate change on poor nations and poor people in the U.S. are enormous."[146]

In response to these pressures, and acting on its rhetoric from the 1992 campaign, in mid-1996 the Clinton administration reversed the U.S. government's long-standing opposition to binding international measures to reduce emissions of greenhouse gases. At the second conference of the parties to the Climate Change Convention, Under Secretary of State for Global Affairs Timothy Wirth announced that the United States would support negotiation by the end of 1997 of a binding international agreement that would require parties to the treaty to undertake such reductions. In sharp contrast to past U.S. government policy, especially during the Bush administration, Wirth expressed strong support for the second assessment report of the IPCC, agreeing that climate change was underway and likely caused, at least in part, by industrial and other human activities. Thus, the Clinton administration endorsed the view that greenhouse gas emissions, especially carbon dioxide, were probably already warming the earth, and through Wirth it urged other countries to join it in 1997 to negotiate an agreement with "realistic, verifiable and binding" greenhouse gas emissions targets. In an interview Wirth said that by "Saying that we want to have a target that is binding is a clear indication that the United States is very serious about taking steps and leading the rest of the world."[147]

In the change of policy announced by Wirth, the United States did not endorse the steep and immediate cuts in carbon dioxide emissions proposed by the small island states, instead calling for emphasis on market-based strategies like tradable emission permits to persuade industries to cut their emissions.[148] The Clinton administration wanted to announce its policy shift after the 1996 presidential elections, but the diplomatic timetable would not permit it to do so.[149] Wirth's position was probably intended in part to deflect election-year criticism from industries opposed to cutbacks in greenhouse gas emissions. Its decision to put off negotiation of specific targets to 1997 was an effort not to "disrupt this year's presidential election in November," according to a press report.[150] One reporter said that "although the Clinton administration has devised no specific target or timetable, its position signals a more aggressive campaign against climate change than the voluntary measures it endorsed in the past, and puts it on a course that promises to be politically difficult."[151]

As anticipated, the Global Climate Coalition (GCC), made up of industries opposed to greenhouse gas emissions reductions (primarily the fossil fuel and automobile industries), attacked the Clinton administration for being too sympathetic to cuts in U.S. greenhouse gas emissions. The organization issued a document signed by 100 U.S. companies asking for no action on climate change. A very small minority of scientists financed by the fossil fuel industry (e.g., S. Fred Singer and other participants in the dubious Science and Environmental Policy Project) remained critical of the IPCC and its new findings. They had lost most of their credibility with the scientific community at large and with the Clinton administration, but they were still listened to by many Republican members of Congress. In a counter-attack, Wirth said that the GCC was "bent upon destroying" the Climate Change Convention.[152] He dismissed the GCC and other critics of the administration's new policy by calling them "special interests bent on belittling, attacking and obfuscating climate change science."[153]

The White House apparently decided to "gamble" on opinion polls that showed that voters wanted action on the environment and that there was "a new mood of optimism among some industrialists about the opportunities that climate might bring."[154] As Hatch showed in the case of Bush administration's climate change policy and politics, the pluralist nature of the American political system and foreign policy-making process, permitted many disparate actors to access the process and thereby pressure the administration. Similarly, in the Clinton administration, many actors pressed their views, but this time the environmentalists and other advocates of equity-promoting policies had sympathizers in the White House and foreign policy bureaucracy and, most important, many business interests supported binding targets for greenhouse gas emissions reductions. While opposition remained, the array of forces was—like during the Bush period—largely favorable to the goals of the administration in power. But the Clinton administration did not go further, fearing that opposition forces would exploit the U.S. policy shift during the 1996 presidential election.

As shown in chapter 5, in the run up to the 1997 Kyoto climate change conference, the Clinton administration was confronted by the Senate's Byrd-Hagel Resolution. That resolution sent the message to the Clinton administration that it could not take bold steps in agreeing to a protocol to the Climate Change Convention that were substantially more robust in implementing international environmental equity. Hence the Clinton team went to Kyoto hoping to prevent an agreement with substantial new commitments for the United States. In the event, however—after the intervention of Vice President Gore—the United States agreed to the Kyoto Protocol, which required the United States to reduce its greenhouse gas emissions by seven percent by 2012 while not requiring developing countries to take on any emissions limitations or reductions. The United States was able to persuade other developed countries to agree to so called flexible mechanisms (e.g., emissions trading, the use of carbon sinks like forest and farmland, etc.), including the Clean Development Mechanism. In subsequent climate change talks, the United States was viewed by other countries—including other developed countries, notably the West Europeans—as an obstacle to more forthright action based on its responsibility for climate change, particularly at the sixth conference of the parties in November 2000. This was attributable to the still strong influence of businesses and members of Congress opposed to U.S. action on climate change. But, again, the Clinton administration diplomats—while preventing agreement at COP-6—were not saying that the United States had no responsibility for its greenhouse gas emissions. They were instead trying to find a compromise between now-accepted obligations (toward international environmental equity in the context of climate change) and the political realities at home.

Similar circumstances obtained with regard to the Convention on Biological Diversity. American and foreign corporations with influence in the United States, including Merck and Genentech, joined with the Biotechnology Industry Organization (a trade association with over 500 member companies, academic institutions, and other organizations), agricultural groups (including the U.S. Council for International Business, the American Seed Trade Association, Archer Daniels Midland Company, and the American Corn Growers Association), and environmental organizations (e.g., the World Resources Institute, the World Wildlife Fund, and the Environmental and Energy Study Institute) to pressure the Clinton administration to sign the Biodiversity Convention.[155] Reflecting the new U.S. policy toward biodiversity, Under Secretary Wirth said that the "vast wealth of genetic information is critical to our long-term economic and environmental integrity, and we must do all that we can to preserve it. . . . The [twenty-first] century will surely be the century of biology, and we must be engaged in order to fully utilize these remarkable opportunities for new sources of food, fuel, fiber, and pharmaceuticals."[156] The Clinton administration policy was that the Convention presented "a unique opportunity for nations not only to conserve the world's biological diversity but also to realize economic benefits from the conservation and sustainable use of its genetic resources."[157] Working in close consultation with the pharmaceutical and biotechnology industries, as well as environmental groups, the Clinton administration crafted ways for the United States to join the Biodiversity Convention.[158] In transmitting the convention to the Senate, the administration asked that seven "understandings" be included in the resolution of advice and consent to ratification, thus delimiting interpretation of the convention from the U.S. perspective.[159]

Here too, the access of politically active actors resulted in a change in U.S. posture that was at least marginally more accepting of international environmental equity and more closely resembled the emerging international consensus on international environmental equity. Many actors remained opposed to U.S. accession to the biodiversity treaty, including conservative think tanks and conservative members of Congress, preventing ratification of the treaty by the U.S. Senate. Nevertheless, the United States signed the treaty and began acting in accordance with its provisions insofar as U.S. law allowed. Most important from the standpoint of equity, the Clinton administration and the foreign policy bureaucracy accepted that developing countries should share in the fruits of biotechnology development when their genetic resources are used, and that those countries ought not be unfairly exploited by industries utilizing those resources.

Conclusion

The U.S. government during most of President Bush's tenure did not view equitable sharing of the benefits, burdens and decision making authority associated with UNCED agreements and treaties as being in either the administration's or the nation's interest. Powerful business interests and influential individuals, most notably White House chief of staff Sununu, were able to prevent the U.S. government from being more accepting of international environmental equity. Nevertheless, in the last two-thirds of 1992, the Bush administration either succumbed to the demands of other interests sympathetic to equity considerations or at least failed to prevent such considerations from becoming part of the UNCED agreements and conventions, as well as the early follow-on negotiations. Indeed, it was pressure from scientists, environmentalists, and the media that persuaded the Bush White House to engage at all in the climate change negotiations.[160] Thus, despite much opposition from the U.S. government under President Bush, the Rio Declaration, Agenda 21 and the treaties agreed at Rio were signed by most of the world's countries, setting precedents that were to be incrementally advanced by the Clinton administration and other governments.

The Clinton administration was influenced by most of the same actors that pressed the Bush administration. But the coalitions of those actors, their orientations toward international agreements that contained equity provisions, and the sympathetic reception given to many of them by the administration, led to different results. Environmental and development organizations continued to press the government, but with the Clinton administration they received a serious hearing, which was not surprising considering that many former officials of those organizations joined the administration. This alone would not have been enough. After all, Bush EPA administrator and head of the U.S. Earth Summit delegation William Reilly was likewise an environmentalist and tried to persuade the Bush administration to be more forthcoming on UNCED equity issues. There was also Vice President Gore, very influential in the White House, for whom environmental protection and international environmental equity were deep personal commitments. In addition, there was a strong emergent coalition of businesses, environmentalists, scientists, and other actors that saw U.S. participation in the UNCED agreements and treaties, as well as follow-on negotiations and measures, as favorable to U.S. interests or to their own interests and objectives.

Both the Bush and Clinton administrations developed and defended their policies in an overall international context of the emerging consensus with regard to international environmental equity. While the Bush administration generally and the Clinton administration occasionally were not enthusiastic about joining that consensus, they could not avoid it. They were compelled to engage various governmental and nongovernmental actors in discussions about the merits of international equity as an objective of global environmental policy. Many countries, primarily the developing states and several countries from Western and Northern Europe, pressed the United States to be more forthcoming with regard to new funding, technology transfers, democratization of funding agencies, and sharing the burdens of efforts to prevent ozone depletion, climate change, and other manifestations of widespread environmental pollution. United Nations organs and international environmental institutions continued to act as the forums for international environmental deliberations. The IPCC, especially with its second assessment report, bolstered the argument that human activities and especially industrialization were contributing to potentially devastating changes to the earth's climate. Thus, although domestic influences were much more important, there were international influences that contributed at least indirectly to the U.S. government's acceptance of international equity as an objective of global environmental policy.

Why was the U.S. government so slow to accept international environmental equity as a policy objective? In trying to explain the Bush administration's policy compared with that of the Western Europeans, Brenton suggests that there are many explanations, including the challenge from the right wing of the Republican party mentioned above. According to Brenton:

> There was, however, also a larger factor at work which has implications for international environmental cooperation in the future. The EC nations, by virtue of their size and proximity, are by now well adjusted and attuned to doing environmental business by international negotiation. They have learned to be attentive to international currents of opinion and to be ready to look for the compromise necessary to get an agreed result. The U.S. has not yet learnt this lesson to anything like the same degree. Its environmental policy-making is much more exclusively focused on domestic politics than is true for any state in Europe.[161]

The evidence, for the most part, supports Philip Shabecoff's analysis of the importance of domestic politics: "The posture of the United States during the summit negotiations was a textbook illustration that the *realpolitik* that motivates participants in international negotiations is not necessarily or even usually the interests of their nation. Their positions are frequently driven instead by the narrow and immediate partisan needs of whoever is in power."[162] During the Clinton administration, the array of domestic actors trying to influence the U.S. government changed, as did the goals and interests of the White House. Domestic actors in large measure determined the U.S. government's posture on international environmental equity during both the Bush and Clinton administrations. Those domestic forces contributed to a significant shift in policy that was more attuned to the emerging consensus on international environmental equity, but domestic forces also prevented the Clinton administration from implementing the new policy in a major way.

Notes

1   United Nations, *Report of the United Nations Conference on Environment and Development, Volume III: Statements Made by Heads of State or Government at the Summit Segment of the Conference* (New York: United Nations, 1993) [A/CONF.151/26/Rev.1 (Vol.III)], pp. 77-78.

2   Those states were also viewed as sympathetic to the pro-business rhetoric of independent presidential candidate Ross Perot. Michael Wines, "Bush and Rio: President Has an Uncomfortable New Role In Taking Hard Line at the Earth Summit," *New York Times*, 11 June 1992, p. A12.

3   Heritage Foundation, *Guidelines for the UN Environmental Conference: A United Nations Assessment Project Study* (Washington, DC: Heritage Foundation, 1991), pp. 1-2.

4   Russell S. Frye, "Uncle Sam at UNCED," *Environmental Policy and Law* 22, 5/6 (1992), p. 343.

5   Marvin Soroos, "From Stockholm to Rio: The Evolution of Global Environmental Governance," in Norman J. Vig and Michael E. Kraft, *Environmental Policy in the 1990s* (Washington: CQ Press, 1994), p. 317.

6   Michael Wines, "Bush Likely to Go to Ecology Talks," *New York Times*, 7 May 1992, p. A17.

7   Norman J. Vig, "Presidential Leadership and the Environment: From Reagan and Bush to Clinton," in Vig and Kraft, p. 87.

8   According to the U.S. UNCED coordinator, Robert Ryan, "we went to unbelievable lengths during the UNCED to consult with the Congress and people on all sides, get their views; we had Congressmen on the delegations." Robert Ryan, Jr., former Director, U.S. Office of the United Nations Conference on Environment and Development, Department of State, telephone interview by author, 12 January 1996.

9   Lawrence E. Susskind, *Environmental Diplomacy: Negotiating More Effective Global Agreements* (New York: Oxford University Press, 1994), p. 39.

10   U.S., Congress, House, *Concurrent Resolution 263*, 102nd Cong., 1st sess., 3 January 1992.

11   Letter from Members of Congress to President George Bush, 21 February 1992, reprinted in U.S., House, Joint Hearings, Committee on Foreign Affairs and Committee on Merchant Marines and Fisheries, testimony of Curtis Bohlen, 26 February 1992, *U.S. Policy Toward the United Nations Conference on Environment and Development*, 102nd Cong., 2nd. sess., 26, 27 February, 21, 28 July 1992, pp. 288-292.

12   Ryan.

13   Congressional Research Service, *International Environment: Briefing Book on Major Selected Issues*, report prepared for the U.S. House of Representatives Committee on Foreign Affairs, July 1993, p. 78.

14   U.S., *U.S. Policy Toward the United Nations Conference on Environment and Development*, statement of Representative Dennis M. Hertel, 21 July 1992, p. 42.

15   U.S., *U.S. Policy Toward the United Nations Conference on Environment and Development*, statement of Representative Dante Fascell, 21 July 1992, p. 46.

16   *Boston Globe*, 30 November 1989, cited in Tony Brenton, *The Greening of Machiavelli* (London: Earthscan Publications, 1994), p. 171.

17   Brenton, p. 172. Concerns in the Bush administration about public attitudes toward the Earth Summit were demonstrated by the announcement of these programs in the President's State of the Union address.

18   Soroos, "From Stockholm to Rio," p. 316.
19   John Holusha, "Poll Finds Skepticism in U.S. About Earth Summit," *New York Times*, 11 June 1992, p. A13.
20   Frye, p. 342.
21   Rose Gutfield, "How Bush Achieved Global Warming Pact with Modest Goals," *Wall Street Journal*, 27 May 1992.
22   Ryan. There was lobbying, nevertheless, on chapters dealing with energy, transport, etc.
23   Mostafa Tolba, director of UNEP, said that the Biodiversity Convention was designed to protect intellectual property rights and industry investments. But Tolba understood that "this [1992] is an election year in the United States and it is not the right time to bring up anything that touches on industry or technology or property rights." Interview with Philip Shabecoff, cited in Philip Shabecoff, *A New Name for Peace: International Environmentalism, Sustainable Development, and Democracy* (Hanover, NH: University Press of New England, 1996), p. 153.
24   William K. Reilly, "Reflections on the Earth Summit," memorandum to all EPA employees, 15 July 1992, reprinted in U.S., *U S. Policy Toward the United Nations Conference on Environment and Development*, p. 397.
25   Michael T. Hatch, "Domestic Politics and International Negotiations: The Politics of Global Warming in the United States," *Journal of Environment and Development* 2, 2 (Summer 1993), p. 23.
26   Michael Grubb et al., *The Earth Summit Agreements: A Guide and Assessment* (London: Earthscan, 1993), p. 37.
27   William Nitze, Alliance to Save Energy, interview cited in Gareth Porter and Janet Welsh Brown, *Global Environmental Politics* (Boulder: Westview Press, 1996), p. 62.
28   See Statement of the National Coal Association on S.324 National Energy Policy Act of 1990 to the Committee on Energy and Natural Resources, U.S. Senate, 20 April 1990; testimony of the Global Climate Coalition on 1990 Draft of S.324, 19 April 1990; testimony of the Global Climate Coalition before the Energy and Power Subcommittee of the Committee on Energy and Commerce, U.S. House of Representatives, 3 March 1992; see also Hatch, pp. 23-25, 36.
29   Steven Greenhouse, "Ecology, the Environment and Bush," *New York Times*, 14 June 1992, p. E1.
30   See Paul Hawken, *The Ecology of Commerce: Doing Good Business* (New York: Harper Business Publications, 1993).
31   Stephan Schmidheiny, *Changing Course: A Global Business Perspective on Development and the Environment* (Cambridge, MA: MIT Press, 1992), pp. 131-32.
32   Hatch, p. 23.
33   The General Assembly resolution establishing UNCED (A/44/228) called for NGO involvement. The UNCED NGO accreditation and participation process is described in Ann Doherty, "The Role of Nongovernmental Organizations in UNCED" in Bertram I. Spector, Gunnar Sjostedt and I. William Zartman, eds., *Negotiating International Regimes: Lessons Learned from UNCED* (London: Graham and Trotman, 1994), pp. 199-218. Practically all those NGOs that applied for accreditation were obliged. The very large number of NGO participants in many ways limited the ability of individual NGOs to influence the negotiations. The accreditation was an "obvious" attempt, according to one report, to limit NGO participation. Editorial, *Environmental Policy and Law* 22, 4 (August 1992), p. 200.

34   This compares with about 130 NGOs at Stockholm, ten percent from developing countries.

35   *Report on the Participation of Non-Governmental Organizations in the Preparatory Process of the United Nations Conference on Environment and Development* (Vienna: Center for Applied Studies in International Negotiations, 1992), p. 5.

36   Grubb et al. declare that "There is little doubt that NGOs made major contributions to the process. . . ." Grubb et al., p. 44.

37   Mathias Finger, "Environmental NGOs in the UNCED Process," in Thomas Princen and Mathias Finger, eds., *Environmental NGOs in World Politics: Linking the Local and the Global* (London: Routledge, 1994), pp. 207-208.

38   UNCED prepcom Working Groups met in three types of forums: (1) formal meetings, where states usually made their official statements, which were open to NGOs; (2) informal meetings (the so-called "formal-informals"), which were translated but not transcribed and where negotiations occurred (access by NGOs was determined by the working group chair); (3) informal-informals, conducted in English, which varied from open-ended meetings in conference rooms (NGOs sometimes permitted) to small meetings held in chairmen's offices (NGOs excluded). Described in "PrepCom: Third Meeting," *Environmental Policy and Law* 21, 5/6 (1991), p. 190.

39   Southern NGOs were especially influential in negotiations on intellectual property rights. Finger, p. 207.

40   Peter M. Haas, Marc A. Levy and Edward A. Parson, "Appraising the Earth Summit: How Should We Judge UNCED's Success?" in Ken Conca et al., eds., *Green Planet Blues: Environmental Politics from Stockholm to Rio* (Boulder: Westview Press, 1995), p. 160.

41   Ibid.

42   CAPE '92's mandate was to "promote urgently needed U.S. leadership at UNCED." Its members were the Environmental Defense Fund, Friends of the Earth, National Audobon Society, National Wildlife Federation, the Natural Resources Defense Council, and the Sierra Club.

43   This was because developed country negotiators had the least flexibility on these issues, which were expected to have potentially substantial impact on their home countries. "Cross-cutting" issues are those that were important for most all aspects of the UNCED deliberations. For example, every chapter of Agenda 21 and both conventions had paragraphs on financing and technology transfer. A failure to resolve these issues would have prevented completion of these agreements.

44   On the order of 90 percent of the NGO proposals made it into these chapters. The NGOs accomplished this through cooperation with delegates from the EC, Canada, Australia and New Zealand. "A Summary of the Proceedings of the Fourth Session of the UNCED Preparatory Committee," *Earth Summit Bulletin* 1 (April 1992), <http://www.iisd.ca/ linkages/>. Timothy Wirth said that the "heroes of the Earth Summit in Rio were not the heads of state but the NGOs who defined and drove Agenda 21." Timothy E. Wirth, "Sustainable Development and National Security," address before the National Press Club, Washington, DC, 12 July 1994, *U.S. Department of State Dispatch* 5, 30 (25 July 1994).

45   Finger, pp. 207-208.

46   Ryan. The environmental NGOs that tried to influence U.S. global environmental policy in the early 1990s included the following: Sierra Club, National Audobon Society, National Wildlife Federation, Environmental Defense Fund, Natural

Resources Defense Council, Greenpeace, Friends of the Earth, Global Tomorrow Coalition, Earth Island Institute, and the Rainforest Action Network.

47   Other countries with NGO personnel in their delegations at some point in the UNCED negotiations included: Australia, Norway, the Netherlands, UK, USSR and subsequently Commonwealth of Independent States, Denmark, Finland, New Zealand, Switzerland, France and India. Finger, p. 208.

48   On more than one occasion Porter included his criticisms in congressional testimony. Importantly, he was critical of the U.S. negotiating briefs for all the prepcoms, but he was not critical of E.U. Curtis Bohlen, Robert Ryan and other key U.S. UNCED diplomats, who he saw as victims of the U.S. negotiating position. See U.S., House of Representatives, Committee on Foreign Affairs, Subcommittee on Western Hemisphere Affairs, statement of Gareth Porter, Director, International Program, Environmental and Energy Study Institute, *The United Nations Conference on Environment and Development*, 102nd Cong. (4 February 1992), 1993, p. 20.

49   Shabecoff, *A New Name for Peace*, p. 150. The chief U.S. UNCED negotiator said the U.S. UNCED delegations "had marvelous input from a great number of NGOs throughout the process." U.S., *U.S. Policy Toward the United Nations Conference on Environment and Development*, testimony of E.U. Curtis Bohlen, 26 February 1992, p. 11.

50   Such a meeting would have been "inconceivable under Sununu." William K. Stevens, "Washington Odd Man Out, May Shift On Climate," *The New York Times*, 18 February 1992, p. C11.

51   U.S., *U.S. Policy Toward the United Nations Conference on Environment and Development*, statement of Robert Engleman, Population Crisis Committee, 21 July 1992, p. 186.

52   Hatch, pp. 19-20. It is important to note that opponents of aggressive U.S. action on climate change also had—and have—exceptional access to the legislative process. The Senate committee controlling energy legislation (Committee on Energy and Natural Resources) was largely made up of Senators from states producing coal, oil and gas, and was chaired by Senator Bennet Johnson from Louisiana (an oil producing state).

53   See, for example, NGO testimony and statements before Congress in U.S., *U.S. Policy Toward the 1992 United Nations Conference on Environment and Development*.

54   Hatch, p. 24.

55   This phenomena outside the specific INC/UNCED context is described in Barbara J. Bramble and Gareth Porter, "Non-Governmental Organizations and the Making of U.S. International Environmental Policy," in Andrew Hurrell and Benedict Kingsbury, eds., *The International Politics of the Environment* (Oxford: Oxford University Press, 1992), pp. 313-53.

56   See Julie Fisher, *The Road from Rio: Sustainable Development and the Nongovernmental Movement in the Third World* (Westport: Praeger, 1993), especially pp. 123-28. This may have contributed to some of the tension that developed between Southern and Northern NGOs at Rio.

57   Atiq Rahman and Annie Roncerel, "A View from the Ground Up," in Irving Mintzer and J. Amber Leonard, eds., *Negotiating Climate Change* (New York: Cambridge University Press, 1994), p. 243. According to one observation, "the non-governmental participants brought a freshness of vision, a concern about equity, and a measure of pragmatism that may not otherwise have been available to the diplomats. . . ." Mintzer and Leonard, p. 322.

58 Epistemic communities are networks of knowledge-based experts "with recognized expertise and competence in a particular domain and an authoritative claim to policy-relevant knowledge within that domain or issue-area." They help states identify their interests, frame the issues for collective debate, propose specific policies, and identify salient points for negotiation. Peter M. Haas, "Introduction: Epistemic Communities and International Policy Coordination," *International Organization* 46, 1 (1992), p. 3.

59 See Richard Elliott Benedick, *Ozone Diplomacy* (Cambridge, MA: Harvard University Press, 1998); Peter M. Haas, *Saving the Mediterranean* (New York: Columbia University Press, 1990); Lynton Keith Caldwell, *Between Two Worlds: Science, the Environmental Movement and Policy Choice* (Cambridge, England: Cambridge University Press, 1992); Karen T. Litfin, *Ozone Discourse: Science and Politics in Global Environmental Cooperation* (New York: Columbia University Press, 1994).

60 Caroline Thomas, *The Environment in International Relations* (London: Royal Institute of International Affairs, 1992), p. 192.

61 See J.T. Houghton, G.J. Jenkins, and J.J. Ephraums, eds., *The IPCC Scientific Assessment* (Cambridge, England: Cambridge University Press, 1990).

62 In diplomatic parlance, "more study" often means "do nothing." But, as the Montreal Protocol process showed, "doing nothing" can allow for developments of overwhelming importance. Discovery of the Antarctic ozone hole and subsequent confirmation of the connection between stratospheric ozone depletion and human activities led to a ban on chlorofluorocarbons and their ilk, thus necessitating the substantive provisions for international equity in the 1990 London amendments, which were meant to foster global participation in the treaty.

63 See, for example, National Academy of Sciences, *Policy Implications of Greenhouse Warming* (Washington: National Academy Press, 1991); Paul C. Stern, Oran R. Young, and Daniel Druckman, eds., *Global Environmental Change* (Washington: National Academy Press, 1992).

64 See, for example, Daniel Bodansky, "Prologue to the Climate Change Convention," in Mintzer and Leonard, pp. 46-49.

65 Brenton, p. 172.

66 Hatch, pp. 1-39.

67 Rio Declaration on Environment and Development, Principle 15; United Nations Framework Convention on Climate Change, Art. 3., Para. 3. The 1990 Houston Economic Declaration of the G-7 was one precursor, albeit a weak one, to the precautionary principle. It pointed out that "In the face of threats of irreversible environmental damage, lack of full scientific certainty is no excuse to postpone actions which are justified in their own right."

68 Interview with Pomerance by Philip Shabecoff, cited in *A New Name for Peace*, p. 152.

69 William A. Nitze, "A Failure of Presidential Leadership," in Mintzer and Leonard, pp. 187-200.

70 Shabecoff, *A New Name for Peace*, p. 137.

71 Susan F. Drake, former advisor to U.S. UNCED delegation and officer in the Office of Environmental Protection, Bureau of Oceans and International Environmental and Scientific Affairs, Department of State, telephone interview by author, 1 November 1995.

72 Chris Thompkins, Minister's Representative, Environmental Protection International, Department of the Environment, United Kingdom, telephone interview by author, 7 August 1996.

73   Hatch, p. 16.

74   Ibid.

75   R.W. Apple, Jr., "Leaders Express Support for Gorbachev," *New York Times*, 17 July 1991), p. A10. At the prepcom, the U.S. distributed brochures listing its environmental accomplishments during the previous year, hoping to avoid being labeled "anti-environment." Ibid., p. A10.

76   Brenton, p. 171.

77   Stanley P. Johnson, *The Earth Summit* (London: Graham and Trotman, 1993), p. 58.

78   International pressure was more successful in other areas. For example, the United States was persuaded by substantial pressure from the other OECD countries to agree to references to "new and additional" resources that permeate the UNCED agreements and conventions. Grubb et al., p. 29. The U.S. was willing to do this provided that there be no mention of specific amounts of aid. Indeed, the term "new and additional" was a significant step toward greater equity. The U.S. had agreed to "new" sources of aid for the Montreal Protocol, but later said that these "new" funds could be diverted from existing aid flows. Thus the push for "new and additional" funds in the UNCED process.

79   See Paul Lewis, "U.S. Under Fire in Talks at UN on Environment," *New York Times*, 24 March 1992.

80   David E. Pitt, "Europe Moves to Aid U.S. in Rio," *Earth Summit Times*, 6 June 1992, p. 1.

81   Alternatively, Japan was described as the world's "first environmental superpower." James Brooke, "U.S. Has a Starring Role at Rio Summit as Villain," *New York Times*, 2 June 1992, p. A10.

82   Scott A. Hajost, "The Role of the United States" in Luigi Campiglio et al., eds., *The Environment After Rio: International Law and Economics* (London: Graham and Trotman, 1994), p. 19.

83   Ernst Ulrich von Weizsacker, *Earth Politics* (London: Zed Books, 1994), p. 170.

84   Marc A. Levy, Robert O. Keohane, and Peter M. Haas, "Improving the Effectiveness of International Environmental Institutions," in Peter M. Haas, Robert O. Keohane and Marc A. Levy, *Institutions for the Earth* (Cambridge, MA: MIT Press, 1993), pp. 397-426, especially p. 406.

85   Gunnar Sjostedt, Bertram I. Spector, and I. William Zartman, "The Dynamics of Regime-building Negotiations" in Bertram I. Spector, Gunnar Sjostedt and I. William Zartman, eds., *Negotiating International Regimes: Lessons Learned from UNCED* (London: Graham and Trotman, 1994), p. 18.

86   Shabecoff, *A New Name for Peace*, p. 139.

87   Hajost, p. 18.

88   Ibid., p. 19. The U.S. had previously joined with Saudi Arabia in trying to block progress on climate change negotiations.

89   The treaty was weak, to be sure, but the Bush administration knew that the Vienna Convention, which led to a strong ozone regime, was even weaker. Members of the administration with any knowledge of environmental negotiations had to recognize that a precedent was being set. Perhaps that is precisely the message that was being sent after Sununu's sacking—and in the run-up to the presidential elections.

90   See Hatch, pp. 1-39.

91   Ibid., p. 10.

92   Or so it was said at the time. Then Senator Albert Gore suggested a different explanation: "Though I count Secretary Baker as a close friend and hold him in the

highest regard, one cannot help but wonder whether his recusal on climate change—which hasn't been matched by a withdrawal from discussions about our policy toward OPEC, the Gulf crises, or other issues with a direct impact on oil companies—had something to do with his keen political sense that he was never going to win the argument with Sununu, and he did not want to be associated with the disastrous and immoral policy on which the White House has been insisting." Albert Gore, *Earth in the Balance: Ecology and the Human Spirit* (New York: Houghton Mifflin, 1992), p. 175. Secretary of State Baker had a habit of disassociating himself from "issues that might backfire." James A. Nathan and James K. Oliver, *Foreign Policy Making and the American Political System* (Baltimore: The Johns Hopkins University Press, 1994), p. 66.

93   Specific targets and timetables were first discussed at a 1988 Toronto conference that called for 20 percent reductions in carbon dioxide emissions by 2000. See United Nations Environment Program, *Proceedings of the World Conference on the Changing Atmosphere: Implications for Global Security* [Toronto, 27-30 June 1988] (Nairobi: UNEP, 1988). The U.S. did not accept the conclusions of the Toronto conference; it was opposed to targets and timetables.

94   Hatch, pp. 10-15.

95   This was in part a consequence of the Bush administration's knee-jerk opposition to any policy that smelled of central planning.

96   Alan D. Hecht and Dennis Tirpak, "Framework Agreement on Climate Change: A Scientific and Policy History," *Climatic Change* 29 (1995 [draft of 18 October 1994]), p. 391. Hecht and Tirpak (at note 48, p. 402) attribute this change to three factors: (1) shifts in senior advisors to the president; (2) persistence of the EPA administrator and other senior officials; (3) a realistic evaluation of what it would take to conclude the climate change negotiations in time for a Rio signing.

97   "U.S. Statements on Commitments," INC V, New York City, 27 February 1992.

98   Stevens, "Washington Odd Man Out, May Shift On Climate," p. C11.

99   See Framework Convention on Climate Change, Art. 4, para. 2, for the final wording. The British negotiators were led by Environment Minister Michael Howard. The compromise wording agreed by the British and Americans was adopted by the subsequent May 1992 INC. Hatch, p. 26-27. The U.S. opposed *binding* targets and timetables because (1) population growth, world fuel prices, and economic growth might push U.S. emissions much higher; (2) a binding commitment to stabilize emissions might require policy actions with unforeseen economic consequences; and (3) the EC position was perceived to be far too optimistic. Hecht and Tirpak, p. 392.

100 Hatch, p. 27.

101 Ibid., pp. 24-30.

102 Ryan.

103 Ibid.

104 Quoted in Gore, *Earth in the Balance*, p. 176.

105 Japan and Canada often joined the U.S. in its unwillingness to discuss new and additional funds for the South. "All Mouth, Too Little Money," *ECO* (16 September 1991), cited in Thomas, p. 186.

106 Hecht and Tirpak, p. 388. Hecht and Tirpak continue: "Remember that the negotiations began after a period of eight years during which the U.S. was largely perceived as a reluctant partner in international issues and had denied that climate change was an important problem. It also came at a time of great euphoria in Europe. Communist governments in many countries were losing power. The EC was moving

quickly toward monetary and social union. This excitement carried over to environmental policy as many European ministers vied to make Europe the center of influence on environmental issues. These ministers also saw targets and timetables as a means of motivating their own governments (especially trade and finance ministries) to move policies in a more green direction." See also Brian Wynne, "Implementation of Greenhouse Gas Reductions in the European Community: Institutional and Cultural Factors," *Global Environmental Change Report* (March 1993), p. 102. The negotiating positions of the major groups of states in the climate change deliberations are summarized in Daniel Bodansky, "Draft Convention on Climate Change," *Environmental Policy and Law* 22, 1 (1992), pp. 5-15.

107 Hatch, p. 28.

108 See the Bush administration's policy paper, "U.S. Views on Global Climate Change."

109 "No regrets" policies are those that are justified by beneficial environmental (usually local/regional) or economic effects separate from their impact on climate change. "Comprehensive approach" policies are those that allow all greenhouse gases to be incorporated into emissions, not just carbon dioxide (the most damaging pollutant over all and the one most closely tied to developed economies). The Bush administration had insisted that CFCs—believed to be potent greenhouse gases that were already being reduced in the context of the Montreal Protocol—be included. It dropped this requirement after it became clear that the overall effect of CFCs on climate change was not as important as initially thought. See Hecht and Tirpak, especially pp. 373, 386.

110 Hatch, p. 29. During the election the issue of carbon dioxide emissions generated great political interest. Senator Al Gore, the Democratic vice presidential candidate introduced a bill (S2668) to implement the original intent of the FCCC, namely stabilization of carbon dioxide emissions at 1990 levels by 2000. Frye, p. 345.

111 The Bush administration was in general opposed to government programs (especially regulations) to limit greenhouse gas emissions. In contrast, the Clinton administration preferred a more corporatist approach in which government and industry would join in overcoming market barriers that might prevent the use of new technologies that help limit greenhouse gas emissions. See Hecht and Tirpak.

112 During Senate debate in October 1992, vice presidential candidate Gore called on parties to the Climate Change Convention to negotiate a protocol with "effective and binding commitments to action." See Hatch, note 60, p. 38.

113 Hatch, p. 29.

114 Ibid., pp. 30-31.

115 Ibid., p. 31.

116 Marian A.L. Miller, *The Third World in Global Environmental Politics* (Boulder: Lynne Rienner, 1995), p. 109.

117 Brenton, p. 198.

118 Miller, pp. 110-12.

119 Brenton, p. 198.

120 Frye, p. 344.

121 "Rio Conference on Environment and Development," *Environmental Policy and Law* 22, 4 (1992), p. 207.

122 Frye, p. 343.

123 Ryan.

124 "Earth Summit: Bush to Go; White House Still 'Recalcitrant,'" *Greenwire* 7 May 1982, p. 9, cited in Shabecoff, *A New Name for Peace*, p. 153.

125   Shabecoff, *A New Name for Peace*, p. 162.
126   Hatch, pp. 30, 37.
127   Alan E. Boyle, "The Convention on Biological Diversity," in Luigi Campiglio, et al., eds., *The Environment After Rio: International Law and Economics* (London: Graham and Trotman, 1994), pp. 114 and 125.
128   John E. Reilly, ed., *American Public Opinion and U.S. Foreign Policy 1995* (Chicago: Chicago Council on Foreign Relations, 1995).
129   Ibid.
130   Shabecoff, *A New Name for Peace*, p. 65.
131   The effect that U.S. NGO lobbying had on the Clinton administration is described in Gareth Porter and Janet Welsh Brown, *Global Environmental Politics* (Boulder: Westview Press, 1996), pp. 143-44.
132   For examples of criticisms of the World Bank (and the International Monetary Fund, which has made fewer environmental reforms than the World Bank), see Giovanni Andrea Cornia, Richard Jolly, and Frances Steward, eds., *Adjustment With a Human Face* (New York: Oxford University Press, 1989); David Reed, ed., *Structural Adjustment and the Environment* (Boulder: Westview Press, 1992); Oxfam, *A Case for Reform: Fifty Years of the IMF and World Bank* (Oxford: Oxfam, 1995); Carol Capps, Church World Service and Lutheran World Relief, "The International Development Association: Flawed But Essential," statement before the Subcommittee on International Development, Finance, Trade, and Monetary Policy and the Banking Committee, U.S. House of Representatives, 5 May 1993. For the preferred shape of funding mechanisms according to some NGOs, see also Jo Marie Griesgraber, ed., *Rethinking Bretton Woods: Toward Equitable, Sustainable, and Participatory Development* (Washington: Center for Concern, 1994). Environmental and developmental reforms at the World Bank are described in World Bank, *Making Development Sustainable: The World Bank Group and the Environment* (Washington: The World Bank, 1993) and World Bank, *Implementing the World Bank's Strategy to Reduce Poverty: Progress and Challenges* (Washington: The World Bank, 1993).
133   "IPCC Working Group I 1995 Summary for Policymakers," 2 December 1995, facsimile. See J.J. Houghton et al., eds., *Climate Change 1995: The Science of Climate Change* (New York: Cambridge University Press, 1996). See also "UN Framework Convention on Climate Change," *Earth Negotiations Bulletin*, 12, 25 (18 December 1995); William K. Stevens, "Experts Confirm Human Role in Global Warming," *New York Times*, 10 September 1995, pp. A1, A7; Stevens, "Scientists Say Earth's Warming Could Set Off Wide Disruptions," *New York Times*, 18 September 1995, pp. A1, A8; Stevens, "U.N. Warns Against Delay in Cutting Carbon Dioxide," *New York Times*, 25 October 1995, p. A13; Robert C. Cowen, "Global Warming Is Real Many Scientists Agree," *Christian Science Monitor*, 10 October 1995, p. 14.
134   Shabecoff, *A New Name for Peace*, p. 152.
135   According to Shabecoff, "The appointment of Gus Speth as head of the UN Development Program, traditionally a post reserved for an American, was a direct result of the changed administration in Washington." Ibid., p. 181.
136   Chirac cited U.S. ODA as $9 billion; the EU figure, $30 billion.
137   For dedicated analyses of U.S. climate change policy, see the contributions to Paul G. Harris, ed., *Climate Change and American Foreign Policy* (New York: St. Martin's Press, 2000).
138   Albert Gore, "Keynote Address by U.S. Vice-President Al Gore," Commission on Sustainable Development, New York, 14 June 1993, reprinted in *Environmental*

Policy and Law 23, 3/4 (1993), pp. 183-85. The U.S. position to reduce greenhouse gases to 1990 levels by 2000 was reaffirmed in Strobe Talbott, "Sustainable Development of Small Island Developing States," address before the UN Global Conference on the Sustainable Development of Small Island States, Bridgetown, Barbados, 5 May 1994, *U.S. Department of State Dispatch* 5, 23 (6 June 1994).

139 Houghton et al., *Climate Change 1995*.

140 "Constructive Industry Hits INC 10," *ECO* (Climate Negotiations, Geneva), 26 August 1994; "Business Leaders Go Green," *ECO* (Climate Negotiations, Berlin), 31 March 1995; "Industry Split Widens," *ECO* (Climate Negotiations, Berlin), 4 April 1995 (*ECO* from <mlist.ecix1@conf.igc.apc.org>).

141 Owen Bennet Jones, "Insurance Firms Join Alarm Over Global Warming," *Guardian*, 10 July 1996, p. 13.

142 David Barren, Morning Edition, National Public Radio, 19 January 1996. In July 1996, Insurance Initiative, a group of 60 insurance companies worldwide, called on the Parties to the Climate Change Convention to take urgent measures to check greenhouse gas emissions. "Insurance Executives Call for Reductions in Greenhouse Gases," *Daily Report for Executives*, Bureau of National Affairs, 10 July 1996, p. 132.

143 Porter and Brown (1996), p. 60.

144 Figure cited in Paul Brown, "Can the White House Save the World?" *Guardian*, 13 July 1996, p. 18.

145 Companies and associations that were members of the U.S. Business Council for Sustainable Energy Futures included American Standard, Honeywell, Southern California Gas Company, Brooklyn Union, Enron, Energy Standard Devices, North American Insulation Manufacturers Association and the Polyisocyanurate Insulation Manufacturer's Association. "Rival Industry Groups Sharply Differ Over U.N. Global Warming Report," Bureau of National Affairs *Daily Report for Executives*, 17 July 1996.

146 "Churches Launch Campaign to Sway U.S. to Accept Emission Limits," *BNA [Bureau of National Affairs] Environment Daily*, 12 July 1996.

147 John H. Cushman, Jr., "In Shift, U.S. Will Seek Binding World Pact to Combat Global Warming," *New York Times*, 17 July 1996, p. A6.

148 Nevertheless, NGOs, diplomats and commentators said the change in the U.S. position was exactly what the climate change negotiations needed in order to move toward a binding emissions protocol. According to an Agence France Presse report, Wirth's proposal "could ease the deadlock and pave the way for a consensus." Agence France Presse, wire story via Environmental Media Services, "Summit Opens on Curbing Greenhouse Gases," 17 July 1996.

149 Cushman, "In Shift, U.S. Will Seek Binding World Pact to Combat Global Warming," p. A6.

150 Andrew McCathie, "Ministers Face Adoption of Tough Conference Declaration," *Deutsche Presse-Agenter*, 16 July 1996 (trans. Environmental Media Services).

151 John H. Cushman, Jr., "Washington Targets Global Warming," *International Herald Tribune*, 18 July 1996, p. 10. The voluntary measures already adopted by the Clinton administration reduced the growth of greenhouse gas emissions but did not actually reduce overall emissions.

152 Cushman, "In Shift, U.S. Will Seek Binding World Pact to Combat Global Warming," p. A6.

153 Marcus Kabel, "U.S. Injects New Energy into Global Warming Fight," *Reuter European Community Report*, 17 July 1996.

154 Paul Brown, p. 18.

155 See *International Environment Reporter* (2 June 1993), p. 416.

156 Wirth, "Sustainable Development and National Security."

157 "Global Environmental Issues," *U.S. Department of State Dispatch* Supplement 5, 8 (September 1994).

158 Senator Claiborne Pell, "Convention on Biological Diversity," statement before the Senate, *Congressional Record* 140, 142 (4 October 1994), p. 1.

159 Addressing concerns in government and industry that the Biodiversity Convention's wording could result in several interpretations, an interpretive statement was drafted by the administration and members of an industry-environmentalist alliance. The United States now interpreted the provisions of the Biodiversity Convention dealing with intellectual property rights and the financial mechanism to be compatible with U.S. interests. See U.S., Congress, Senate, *Convention on Biological Diversity: Message from the President of the United States*, Treaty Doc. No. 103-20, 20 November 1993 (Washington: U.S. Government Printing Office, 1993).

160 Shabecoff, *A New Name for Peace*, p. 152.

161 Brenton, p. 235.

162 Shabecoff, *A New Name for Peace*, p. 136.

8 International Environmental Equity and Altruistic Principles

The previous chapters argued that consensus has been emerging in the U.S. policy community that environmental threats to U.S. interests require international cooperation if they are to be effectively addressed, and that a credible response to the international equity demands of the developing countries is required for the needed cooperation to materialize. A range of politically active actors, from individuals to environment and development organizations to industry lobby groups, have pressured the government to take these demands seriously—or to avoid doing so. This constellation of perceived tangible environmental threats and international and domestic pressures have contributed to the government's gradual acceptance of international equity as an objective of the global environmental policy of the United States, with various political factors preventing a more robust movement in this direction.

While these considerations provide the dominant explanation for the U.S. government's posture, they are not sufficient. Another explanation must also be considered: international altruism—the commitment to better the lot of the world's less fortunate peoples—which, with respect to global environmental policy, translates into international equity as a value in itself. That is, some people viewed international environmental equity as not merely an instrument for promoting U.S. national interests, but also saw it as a valuable goal *per se*. These sentiments were and are even stronger in some other developed countries. Concrete proof for this is difficult to gather because it frequently requires determining the motivations of policymakers. Nevertheless, this chapter tries to make a case for such a conclusion.

Evidence of Altruistic Motives and Their Influence

The testimony and actions of Americans involved in the UNCED deliberations reveal a significant pressure of altruistic considerations. The nature of global environmental problems themselves contributed to this situation. Thus, the former director of the U.S. Department of State's Office of Environmental Protection observed:

> While it is the job of an international negotiator to protect and advance his own country's interests first and foremost, on questions such as stratospheric ozone depletion or climate change, these interests (e.g., human health, environmental protection) are best served when nations act in concert to address common threats. Put another way, the question of protecting one's "own" climate or ozone layer simply does not arise. Since a depleted ozone layer, for example, has serious implications worldwide, officials negotiating international agreements to address this problem tend by definition to take a broader, more altruistic view than a purely "national" one.[1]

According to Philip Shabecoff, a close observer of the UNCED process (he was given full access to the negotiations by UNCED Secretary General Maurice Strong), the U.S. delegations to the UNCED deliberations included several "talented and dedicated public servants who personally believed in the need for global cooperation to create a new system for protecting the environment and promoting equity. Buff Bohlen [chief U.S. UNCED negotiator], for example, was an officer of the World Wildlife Fund [vice president for policy] before returning to the State Department [in June 1990]."[2]

Some U.S. diplomats involved in the UNCED negotiations indicated more than rhetorical sympathy for people living in poverty in the developing countries. They recognized that U.S. policy had a great impact on those countries and that the United States had a responsibility to lead in international environmental negotiations and agreements. In the early stages of the UNCED negotiations—at least from 1989 to the second preparatory meeting in March 1991—these diplomats had more influence because the Bush administration's hard line view of the Earth Summit had not yet solidified. Their concerns for international welfare found their way into the policy process, and the official U.S. position was, albeit briefly, sympathetic to international equity in the context of UNCED. This proved to be extremely important, however: In later negotiations developing countries were able to invoke the early U.S. acknowledgment of the needs of the poorest countries.[3] As Lumsdaine observes, "Concern for justice can shift power in society toward the powerless, sometimes with the consent of the powerful who have come to acknowledge the demands of justice."[4] This is precisely what happened.

Relatively low-level, altruistically motivated individuals in the Bush administration were able to influence major U.S. positions early in the UNCED process (before early 1992), as well as less salient issues right through to the Earth Summit. This was possible because there was limited interest in UNCED at higher levels in the Bush administration (i.e., the White House) and there was a general lack of bureaucratic coordination with regard to UNCED. This was truer than in other international negotiations due to the monumental scope of the UNCED deliberations and the complexity of the environmental issues being discussed. Suggesting these difficulties, Elliot Richardson described the UNCED process to Congress this way:

> When you add to [complexity and uncertainty of global environmental issues] the unwieldiness of the processes of the U.S. Government and the lack of any clear focus toward UNCED, I think you can account for the slowness with which the administration had developed a clear-cut set of positions [regarding UNCED], and the lag in its assertion of global leadership. . . . the State Department obviously has a major role and has a bureau of its own addressing the environment and to science, when you consider that the Department of Commerce obviously has major interests, the Department of Energy, Department of Interior, Department of Defense, the result is a very difficult bureaucratic exercise. . . . I myself have urged for a long time that the administration focus responsibility at a higher level than it has, and I think part of the problem to date, and perhaps it will not be solved between now and 1992, is that the centers of concern and responsibility in the administration are scattered among a lot of people of coordinate status, all below the level of a cabinet secretary or deputy.[5]

To make matters worse, many of the items being discussed in the UNCED negotiations, especially Agenda 21, were "so complicated that upper reaches of the policy apparatus probably did not understand what was being agreed to."[6]

According to Susan Drake, the U.S. diplomat who helped negotiate the 1989 UN General Assembly Resolution (44/228) that established the framework for UNCED, in the early months of the Bush administration there was little guidance from the State Department on what the U.S. position should be regarding UNCED generally and international equity in the context of UNCED in particular. She had no guidance on how negative the U.S. government would be regarding new funding and technology transfer. Drake confided that White House chief of staff Sununu, who was very much involved in setting U.S. policy on climate change, was not interested in the early negotiations of the Rio Declaration and Agenda 21.[7] Scott Hajost, former head of international affairs at EPA and head of the U.S. delegation to the 1989 UNEP Governing Council that drafted the original proposal for Resolution 44/228, has made similar statements:

> The U.S. was ambivalent at best about convening the 1992 UN Conference on Environment and Development. Early deliberations in the UN were handled by low-level U.S. government officials and with little inter-agency discussion. The first time I was really apprised of the fact that there might be a UN Conference was shortly before I left to attend the 1989 UNEP Governing Council. . . . The U.S. never identified a high-level point person as some of us recommended. UNCED was never made a priority generally nor by U.S. agencies individually, including EPA until a late date. No effective inter-agency process was ever really established until it was probably too late to have a positive impact. No vision was ever articulated on what the U.S. wanted out of the Conference and where it wanted to take it. . . . It refused to address legitimate developing country concerns.[8]

This lack of concern, coordination and understanding on the part of senior policymakers during the early phases of the UNCED deliberations was critical because it permitted altruistically-motivated low- and mid-level officials to influence U.S. policy, and UNCED generally, much more than they could have done had UNCED been a higher priority sooner in the Bush administration. The upshot was that Drake was told to use her own judgment in presenting the U.S. position during UN-sponsored negotiations. She thought that calls for new and additional funds, technology transfers on concessional terms and the like were justifiable and posed no threat to U.S. interests. She did what she thought was, in her words, "fair and equitable."[9] Drake's views on the environment and considerations of international equity derived from her Mennonite background, her "close relationship to the land," and the time she spent (most of her life) living in developing countries (e.g., India and African countries), which gave her an understanding of the problems that people in poor countries face. She recognized that U.S. policies affected developing countries greatly and that the United States, as the only superpower, had a "special responsibility" to those countries.[10]

Thus, in a three-day secret meeting among a few diplomats who wrote Resolution 44/228 (at which she was chairperson), Drake agreed to extensive provisions for international equity in the context of UNCED.[11] When a draft of the resolution was sent to the State Department, Drake was told that the United States could not accept the language on new funds and technology transfer. However, she went over the heads of her State Department superiors to an officer at the National Security Council, telling the NSC that the United States would be the only country not accepting the language of 44/228. The NSC officer was "supportive" of her views, overruling the State Department. This meant that the White House effectively went along with the language that Drake helped draft. Because the Bush administration did not at first have a firm policy on UNCED, Drake said that she was influential in UNCED negotiations from 1989 up to the second prepcom (18 March-5 April 1991). According to Drake, other State Department personnel, including her superiors, were sympathetic to her views and "bought into" the whole UNCED process and thought that the United States had a "responsibility" to UNCED.[12]

Only later, as White House chief of staff Sununu and others began to take an interest in UNCED, did U.S. policy firm up in opposition to equity considerations—and most everything else—in the UNCED deliberations.[13] As noted in the previous chapter, Sununu's nearly preeminent position in the administration on many issues and his control over much of the administration's environmental foreign policy, along with the influence of his "appointees" and other conservative ideologues (e.g., Vice President Quayle) in the White House and executive agencies, contributed to a period in which policies intended to promote international equity were dead-on-arrival—or at least moribund, to be partially revived by others in the Bush administration after Sununu's departure, and enlivened by the Clinton administration. After Sununu left office and other perspectives came to the fore, the U.S. position became less stridently against the South's equity demands. For example, E.U. Curtis Bohlen, head of the U.S. UNCED delegation, said at the fourth UNCED prepcom in New York that "The U.S. accepts that if the world is to fully achieve sustainable development, industrialized countries [presumably including the United States] must generate new and additional financial resources."[14]

Despite the Sununu-driven opposition to international environmental equity, Drake's role in negotiating UN Resolution 44/228 was crucial to Agenda 21 and Rio Declaration negotiations. The resolution formed the basis for all subsequent discussions, most notably in the UNCED prepcoms. The resolution was the outcome of negotiations characterized by North-South debate that culminated in recognition of the Brundtland Commission's linking of environment and development.[15] The equity considerations—the responsibility of the developed countries for current environmental problems, new and additional funds to help developing countries fulfill the UNCED agreements, etc.—in those deliberations and the documents signed at Rio were foreshadowed by Resolution 44/228.[16] Developing countries referred to 44/228 as a victory in codifying their development and equity-related demands. Throughout the UNCED process, the G-77 "endeavored to solidify the issues already contained in Resolution A/44/228."[17] As American opposition to UNCED equity considerations firmed up in 1991, developing countries began to remind the U.S. negotiators that "you agreed before" to the language of Resolution 44/228.[18] The momentum built up by 44/228 was typical of large international negotiations: "Documents are prepared, translated, distributed. You cannot easily jump horses (or change texts) in mid-stream without due warning. Like a heavily-laden supertanker traveling the high seas, it takes more than a quick swing of the wheel to make an intergovernmental conference change direction."[19]

Drake's role attests to the significant influence that low-level people in the Bush administration had on UNCED negotiations.[20] This makes the sources and motivations of their personal and professional perspectives important.

According to the coordinator of U.S. government's UNCED policy and U.S. preparations for the Earth Summit, American diplomats involved in negotiating the Rio Declaration solicited views in symposia from environmental and developmental groups and individuals who "thought about things in equity terms," including philosophers and religious leaders.[21] Diplomats involved in formulating and presenting the U.S. position perceived themselves as negotiating something in which equity considerations were "fairly major."[22] Members of the U.S. prepcom delegations were more aware than the White House of how critical agreement on funding was for the whole UNCED process. One U.S. delegate at the third prepcom said privately, "Unless we can reach agreement on financial issues, UNCED is going to crash."[23] Insofar as U.S. negotiators in the UNCED process followed guidelines from the Bush White House to oppose proposals for new spending on the environment, targets on greenhouse gas emissions, and concessional transfers of technologies, they did so with great reluctance.[24]

Individuals motivated by the idea of equity as intrinsically valuable (among other principled notions, such as the inherent value of environmental protection or general ethical or religious prescriptions) were more active and influential in the Clinton administration than in the Bush administration.[25] Perhaps most important among those individuals was Vice President Gore, who established himself as an environmental activist and as a person concerned about the well-being of poor countries and people well before the advent of Clinton. During the UNCED preparatory process Gore campaigned for U.S. cooperation with other countries to incorporate sustainable development into international environmental deliberations. He attended all the UNCED prepcom meetings and the Earth Summit,[26] where he joined with environmental and development NGOs, sympathetic business groups, and other delegates to press the Bush administration to fulfill developing country demands for new funding and new, more democratic, environmental funding agencies, technology transfer, and commitment by the U.S. government to reduce its carbon dioxide and other greenhouse gas emissions.

Gore wrote a 1992 book, *Earth in the Balance*, in which he described his philosophy toward environment and development issues. In that book he called for a "Global Marshall Plan" to address the world's environment-development crisis, and warned Americans and other developed countries that they would have to lead by example and end their "addiction" to "consumption of the earth itself."[27] "I have come to believe," wrote Gore, "that we must take bold and unequivocal action: we must make the rescue of the environment the central organizing principle for civilization."[28] According to Gore, widespread social injustice would make worldwide acceptance of this principle difficult. Therefore, the "promotion of justice and the protection of the environment must go hand in hand in any society, whether in the context of a nation's domestic policies or in the design of 'North-South' agreements between the industrial nations and the Third World."[29] Gore wrote that "what's required now is a plan that combines large-scale, long-term, carefully targeted financial aid to developing nations, massive efforts to design and then transfer to poor nations the new technologies needed for sustained economic progress, a worldwide program to stabilize world population, and binding commitments by the industrial nations to accelerate their own transition to an environmentally responsible pattern of life."[30] The Global Marshall Plan envisioned by Gore would require, as he told it, "the wealthy nations to allocate money for transferring environmentally helpful technologies to the Third World and to help impoverished nations achieve a stable population and a new pattern of sustainable economic progress."[31] To work, this plan would also require wealthy nations to transition to lower levels of consumption. Specific changes he proposed included reductions in emissions of greenhouse gases, especially carbon dioxide: "it hardly seems reasonable—or ethical—to assume that it is probably all right to keep driving up carbon dioxide. In fact, it is almost certainly *not* all right."[32]

These words were written by an experienced politician who would shortly go on to become the U.S. vice president and who would use his substantial authority between the elections and his taking office to make sure that individuals sympathetic to his views on environment and development were appointed throughout the new government, including the State Department. Indeed, one reason Gore was chosen by then Governor Clinton to be his running mate was Gore's strong environmental record.[33]

Gore's rhetoric was somewhat muted during the campaign, but he generally repeated the themes of his book. Most important, he frequently promoted those themes while he was vice president.[34] According to an assessment of Gore's role in the Clinton administration, Gore argued repeatedly to the president that global environmental issues were, "like Bosnia and other foreign policy challenges, urgent matters not only of national interest but of moral obligation to future generations."[35]

Gore's attention to issues of international environmental equity in practice was consistent with expected relationships between policymakers' value systems and their policy advocacy. If Gore's book and his rhetoric as a Senator were consistent with his true beliefs, his rhetoric while vice president and the actual acceptance by the U.S. government under Clinton of international equity as an objective of global environmental policy were also consistent with those beliefs. The moral relevance of a particular foreign policy issue for a policymaker can lead him to take particular interest in the issue and therefore elevate it on the foreign policy agenda. In addition, the conscience of a policymaker can lead him to decide to support a particular policy action, at least partly because it has moral significance for that policymaker.[36] The appeal of equity itself (and of environmental protection, to be sure) seems to have contributed to Gore's advocacy of, and policy support for, the U.S. government's partial acceptance of international environmental equity as a policy objective.

The contrast between the Clinton and Bush administration policies on this issue fits a historical pattern in the United States. In the post-war period, Democratic administrations have tended to be more "progressive" than Republican administrations in foreign policy; they have been more willing to use tax dollars to finance development in the South, to give economic rather than military aid, to provide grants over loans, to divert resources from bilateral to multilateral aid programs, to seek regional and other international solutions to problems, and to use methods other than military intervention to solve problems in the South that affect U.S. interests.[37] Republican administrations have tended to oppose foreign development aid and they have rarely admitted any moral obligation to help developing countries raise the living standards of their poorest citizens. Democratic administrations have generally advocated "progressive" foreign policies with their rhetoric, and they have tended to put that rhetoric into practice. While they have not accepted U.S. responsibility for poverty in poor countries, "they have tended to advocate helping Third World peoples to improve their standard of living as a moral responsibility of a great and wealthy nation."[38] This historical standard is especially important in light of what is known about Gore and other individuals in the Clinton administration. The Democratic administration, consistent with history, was generally supportive of ideas like international environmental equity, and therefore provided an especially sympathetic and receptive environment in which the views of Gore and others could be entertained.

Polling data from the mid-1990s suggest that, within limits, ordinary Americans were also interested in helping poor people in other countries. A large proportion of Americans agreed in the early 1990s that there were some justifications for foreign aid. Those justifications were, first, humanitarian, followed by environmental and economic rationales. These replaced cold war security concerns as the public's justifications for foreign aid. Many Americans felt a sense of responsibility to help poor countries. The most widely supported kinds of foreign assistance were disaster relief and feeding the hungry and poor. In 1992 a large proportion of Americans (89 percent) agreed that "wherever people are hungry or poor, we ought to do what we can to help them." In polls conducted in 1986 and 1992, a majority of Americans believed that the U.S. government was doing the right amount or less than it should to fight world poverty. In 1993, 72 percent of Americans were in favor of the United States providing humanitarian aid to developing countries. In a poll conducted by ABC news, 70 percent of the respondents supported the United States taking a leading role in providing humanitarian aid to victims of natural disasters or war. Fifty-six percent supported the use of U.S. troops to prevent famine or mass starvation.[39] In a July 1993 poll, a majority of the public (54 percent) said the purpose of U.S. foreign policy should be to "realize human values."[40] A January 1995 poll found that most of the Americans questioned wanted the United States to give some aid to help people in other countries who were in genuine need. Eighty percent agreed that the United States "should be willing to share at least a small portion of its wealth with those in the world who are in greatest need."[41] These general views have prevailed according to more recent polls. It is true, however, that foreign policy elites view foreign aid much more favorably than do the public at large, and while the public do find value in foreign aid, it is not among their top priorities.[42]

As Smillie points out, "Despite a strong sense of 'compassion fatigue' within the international development community, the evidence from dozens of recent and past opinion polls show that the public support for international aid programs has remained consistently and surprisingly high for three decades . . . downward trends (in aid support) are debatable, transitory, or they are simply false. . . ."[43] It is important to emphasize—because it is so rarely acknowledged—that Americans are among those people supporting foreign aid. In a recent comprehensive review of polling data, Kull and Ramsay debunk the notion that Americans do not want to help the world's poor. They show that "an overwhelming majority of Americans embrace the broad principle of giving foreign aid to the needy. . . . *Support derives from both altruism and self-interest.* Majorities embrace the ideas that giving foreign aid helps the U.S. to develop trading partners, preserve the environment, limit population growth, and promote democracy. But overwhelming majorities reject the idea that the U.S. should only give aid when it serves the national interest."[44] Indeed, the opposition to foreign aid that exists among Americans derives from their ignorance of how little is actually spent. When asked how much aid the U.S. government should give to the world, Americans say that the amount should be much higher than it actually is, and they think the United States ought to give the same level of aid as a percentage of GNP as do other countries.[45] These public opinions were evident throughout the 1990s. While most politicians and policymakers failed to acknowledge them, they were arguably known and shared by some people shaping U.S. international environmental policies.

In Lumsdaine's view:

> International influences that support [foreign aid are] not always purehearted and selfless ones. Pressure from the LDCs, fear of seeming stingy, a desire to be part of the group, and a need to assuage troubled consciences jostled with more humane and ethical concerns, as they do in every situation of moral and social suasion. Motives were mixed, and results were mixed, as they always are. But this complex influence arising from countries' sense of their place in international society was not a matter of calculating behavior required to ensure relative power or optimize incentives; it is a process of social influence such as occurs in domestic society and personal life.[46]

It is this international expression of domestic ideas that the next section addresses.

Domestic Welfare Policies and International Development Assistance

The emergence of international environmental equity as an objective of the industrialized countries, and the U.S. government's increasing acceptance of it as an objective of U.S. global environmental policy, are logical extensions of the post-1950 foreign aid regime. Foreign aid programs arose in all the industrialized democracies at about the same time, around the 1950s, and those programs remain in place today. Though the United States was the leading international donor for some time, in recent decades international assistance has become a collective effort of the developed countries, with the United States generally falling behind its contemporaries.[47] Efforts by developed countries to promote international equity through the UNCED process reflected this transformation. The U.S. government's failure to embrace international environmental equity to the extent done so by other developed countries is made clearer when viewed in the context of the foreign aid regime, as is the contention that some individuals involved in the making of U.S. policy valued international equity as a value in itself.

Just as some countries are more generous in their domestic welfare policies than others, some countries are more generous with regard to foreign aid than others. Similarly, just as some people in society are more generous than others, some people in government—including the U.S. government—are more generous than others in the equity-related policies that they support and promote. There is a growing body of research showing a direct correlation between foreign aid and domestic welfare policies of the industrialized democracies.[48] Those developed countries with generous domestic programs of social welfare tend also to have generous foreign aid programs that focus much of their aid on helping poor people in developing countries. Foreign aid is an international institution somewhat analogous to the safety net of domestic welfare programs.[49] Political values and practices that are central components of domestic political institutions are seen to be having a growing impact on, and are reflected in, international politics.[50]

In 1960 Gunnar Myrdal declared that a "welfare world" was needed.[51] A decade later he observed that, at least for Swedes, "*there cannot be any other reason for giving aid than the simple humanitarian impulse to feel solidarity with those who are poor, hungry, diseased, and illiterate, and who meet difficulties in their efforts to rise out of poverty.*"[52] As Myrdal tried to make clear, the very notion that the developed countries should give special consideration to the needs of developing countries and "should even be prepared to feel a collective responsibility for aiding them, is an *entirely new concept dating from after the Second World War.*"[53] As Lumsdaine points out, grants of technological and financial aid to poor countries are an "extremely anomalous" occurrence, at least by the standards of the history of international relations.[54] "Foreign aid constitutes a pivotal, systematic change in the relations between rich and poor countries; a significant revision of the international system on the basis of the recognition of a moral obligation to the weak."[55]

Many converging lines of evidence indicate, according to

Lumsdaine, that the interests of donor states cannot explain all foreign aid. Lumsdaine shows that the main reason for aid by most of the donor countries from 1949 to 1989 was humanitarian concern, although in the U.S. case this is less true.[56] He finds this to be frequently the case regardless of the recipients, the donors and the amount they provided in aid, which groups and politicians in the donor countries were the supporters of aid, public opinion, how aid got started, and how it evolved with time. Foreign aid was motivated by ethical and humane concern about world poverty and from the assumption that world peace and prosperity would be impossible without a generous and just world order in which all people have the opportunity to do well and prosper.[57] Over the first forty years of the foreign aid regime, the largest financial flows from industrialized to developing countries were governed by the principle of beneficence: the affluent should use their wealth to help those in dire poverty living in other countries.[58]

Lumsdaine's study demonstrates three things: (1) Attitudes toward poverty in the development of the social welfare state in the North paved the way for economic assistance to less developed countries. (2) Interactions with other states and of citizens with other people worldwide have influenced countries' aid policies. (3) The principle of help to those in great need implicit in the very idea of foreign aid led to steady modification of aid practices, which focused them more on the needs of the poor and moved them away from the interests of the donors.[59]

Lumsdaine found that there has been a steady change in the foreign aid regime generally:

> There is clear evidence that the OECD countries as a whole, *and also* most of the individual countries of OECD, made steady progress toward: (1) more aid as a percentage of GNP (despite the U.S. decline in this area, which made the OECD total sink during the late sixties and most of the seventies); (2) aid less tightly linked to special, exclusive relationships between a particular donor and recipient; (3) aid more channeled and buffered by multilateral agencies; (4) aid offered at more concessional terms and with fewer conditions on procurement in the donor country; (5) and aid directed more to the poorer nations—particularly the least developed nations of Africa and Asia—and more consciously designed to reach poorer sectors within those (and other) countries.[60]

Lumsdaine pulled apart foreign aid programs and found that they did not look like self-interested behavior:

> Aid funded by DAC [the OECD's Development Assistance Committee]

countries was large-scale, regular, and ongoing; publicly given, internationally monitored, often multilaterally channeled, and generally available to any poor nation; concessionally financed, and usually untied to explicit military, economic, or diplomatic bargains; and provided for projects of economic betterment (usually development rather than relief), designed by technical experts and economists, and subject to formal conditions to assure attainment of humane or developmental purposes or (allegedly) sound economic policies in recipient countries, not to their foreign policies or other external behavior.[61]

In short, much foreign assistance was offered publicly and on concessional terms, and oriented toward development programs.

As the foreign aid regime was established and solidified from the 1960s to the 1980s, donor countries systematically revised their aid programs in ways that made the aid less useful to themselves and more useful to beneficiaries.[62] A growing proportion of foreign aid was given to multilateral institutions like the United Nations Development Program (UNDP) and the World Bank's International Development Association (IDA), which makes loans to very poor countries on highly favorable terms.[63] Aid from all OECD DAC countries going to multilateral institutions increased four times (to 28 percent) between the mid-1950s and the late 1980s. In the U.S. case, multilateral aid increased five times to 20 percent.[64] Increasingly, aid from developed to developing countries was directed toward the poorest sectors in the poorest countries, which were unlikely to offer donors significant economic benefits.[65] As aid became established practice it was increasingly expected of all responsible developed countries.[66] Developed countries have been "concerned about whether they are doing well or poorly by international standards, and there is an attempt to act virtuously in a general way, and not only to fulfill definite, agreed-on obligations clearly monitored."[67] According to a UNDP report, "Since most people approve of aid they do not want their country to be seen as a 'bad donor,' giving proportionately less than other countries."[68]

Considerations of international equity in the UNCED deliberations and provisions for equity in the UNCED agreements and conventions, and particularly many developed countries' support for these considerations, coincided with the evolution of the foreign aid regime described by Lumsdaine and others.

Invoking the models of domestic political systems developed by Esping-Anderson, Noel and Therien classified three types of welfare states: conservative, liberal and social democratic.[69] Conservative welfare states (e.g., Germany) "typically institutionalized corporatist arrangements that espouse existing social categories and preserve them from the challenges posed by capitalist development and new social forces. . . . The growth of welfare represented less a rising commitment to justice than an ever renewed reaction to the challenges of modernity."[70] Liberal welfare states (e.g., the United States) attempt to complement, not counter, market forces by providing individual insurance and means-tested social programs. They tend to "resist intervention, remain suspicious of demands for social support or redistribution, and tend to spend moderately."[71] The social democratic type of welfare state (e.g., Denmark) is most generous in its social welfare programs, viewing such programs as "embodiments of a shared conception of citizenship. In this perspective, social spending appears in a more positive light and, as is the case in Scandinavia, is more likely to be at a high level."[72]

This typology of domestic equity can be roughly extended to international equity. Social democratic welfare states are most likely to have generous foreign aid programs focused on the poorest developing countries, while conservative welfare states tend to be less generous. Noel and Therien show that the values and principles that are central features of social democratic institutions within countries had an effect on the foreign aid regime. Redistribution of financial resources based on need, not just based on markets, is a characteristic both of many domestic societies and international relations. In other words, domestic politics helped shape international politics.[73] Those countries that are generous at home tend to be generous abroad.

The experience of Denmark is instructive.[74] By the 1990s, Denmark was spending 30 percent of its GDP on domestic social welfare and one percent of its GDP on ODA.[75] Its level of international assistance to developing countries was above levels that other developed countries used as long-term goals, reflecting the Danes' "passionate commitment to solving the social problems on the global scene that they have worked to solve at home."[76] The primary objective of Danish development assistance was to alleviate poverty in developing countries, and 45 percent of Danish development assistance was given to multilateral institutions, especially UNDP.[77] One reason that the Danes were so generous was that their comprehensive social welfare system gave them personal security, time and "mental surplus to worry about others than themselves."[78] Danes had, according to UNEP's Lois Jensen, "a highly developed social conscience which includes to help the disadvantaged."[79] Indeed, according to a former high-level member of the U.S. UNCED delegation, not just Denmark but the Nordic peoples generally "think viscerally that helping the poor is a good thing."[80]

Total official development assistance to developing countries in

1992 was $60B,[81] half of what the UNCED secretariat estimated would be needed from the North to implement Agenda 21 programs (and half of what aid would be if all OECD countries adopted and implemented the U.N.'s 0.7 percent official development assistance objective).[82] This amount began falling in the early 1990s. While the level of ODA from OECD countries as a percentage of GNP rose over a two decade period, from an average of 0.33 percent in 1970 to 0.53 percent in 1988,[83] it fell back to about 0.35 percent in 1992.[84] After the 1992 Rio Earth Summit foreign development aid from North to South decreased, both in absolute amounts and in percentage of the donor countries' GNPs.[85]

What explains this? Much of the explanation can be found in the end of the cold war and the end to the developed countries', particularly the United States', need for developing country allies. However, it is worth considering that many of the industrialized countries experienced continuing recessions during this period, only recovering in the late 1990s, and this was accompanied by *reduced provisions for domestic social welfare* during the decade. This resulted in "aid fatigue," an unwillingness to give money to developing countries that were perceived, in any case, as often unable to put the aid they previously received to good use.[86] The Congressional Research Service described the situation succinctly: "In a period of worldwide recessions and tighter financial constraints, how to marshal the resources and deploy them effectively to achieve environmentally sustainable development remains a difficult question."[87] This reduction in development assistance is entirely consistent with the analyses of Lumsdaine and Noel and Therien. The development assistant policies of the developed countries were a reflection of their domestic welfare strategies. Put simply, reductions at home meant reductions abroad.

In contrast, direct foreign private investment in developing countries began to increase substantially in the 1990s, rapidly exceeding ODA. Most of that investment went to a few newly industrializing countries; very little of it reached the poorest countries.[88] Thus, official development assistance remains very important to many developing countries, especially the poorest among them. Therefore, while ODA has been going down lately, much of that aid has been directed to programs designed expressly to help the poorest of the poor. This suggests some level of generosity toward the world's poorest people, a position that is bolstered by other new characteristics of ODA, including greater aid given to multilateral institutions (as opposed to bilateral aid, which gives donors more influence), the high grant-to-loan ratio, and the increasing proportion of aid that is not tied to programs meant to benefit donor countries.[89] Indeed the 1995 UN Copenhagen World Summit for Social Development had as one of its most innovative products the "20/20" compact on human development that calls on developed countries to direct 20 percent of their aid and the developing countries to direct 20 percent of their own budgets to "human priority expenditures" such as basic education and health care, sanitation, clean drinking water, immunization, nutrition and the like.[90] The Social Summit focused on fighting poverty in the least developed countries and was the first time that countries committed themselves to eradicating absolute poverty and to the establishment of strategies and targets for doing so.[91]

In many remarkable ways, the constellation of concern for international equity in the UNCED deliberations mirrored that of the foreign aid regime in the latter half of the twentieth century. Those countries with relatively generous domestic welfare programs tended to be more sympathetic to the broad concerns of developing countries for more equitable relations between North and South. The sub-group of Canada, Australia, New Zealand, and especially the Nordic countries were most sympathetic to the concerns of developing countries.[92] Alternatively, the countries with less generous domestic welfare programs, such as the United States and the United Kingdom, were less willing to meet the equity demands of the developing states. The social welfare states of Europe, especially Scandinavia, struggled against the early hard-line position of the United States on equity considerations, from accepting responsibility for past pollution and willingness to reduce emissions of greenhouse gases to pledges regarding levels of foreign aid and technology transfer. From some perspectives, the negotiations on the Climate Change Convention were, in short, a struggle between the developing countries, the United States, and the other OECD countries, rather than simply the latter two versus the first.[93] At times the other OECD countries as a group were considerably more sympathetic to the developing countries' positions on international equity than that of the United States.

The analyses conducted by Lumsdaine, Noel and Therien, and others shows the United States lags behind most other advanced industrialized countries in its humanitarian aid. This should come as no surprise because their data show a correlation between domestic welfare programs and foreign aid. America's limited concern for international poverty matches its limited concern for poverty in the United States; the percentage of GNP that the United States spends on foreign aid is less than that of almost any other developed country (with the occasional exception of Ireland, one of the poorest OECD countries).[94] The United States is relatively stingy at home; it is also stingy when it comes to foreign aid. Nevertheless, as the foregoing analysis of U.S. UNCED policy showed, there has been a humanitarian component in U.S. global environmental policy. The objectives of some Americans who participated officially in the UNCED deliberations were consistent with the humanitarian aspects of the post-war foreign aid regime, and a substantial percentage of U.S. foreign aid went to the poorest of the poor. Self-interest cannot explain *all* U.S. foreign aid.

A Growing Sense of International Accountability?

Seyom Brown has suggested that there is a growing incongruence between problems encountered by the world's communities and individuals and the capacity of traditional governance systems (dominated by nation-states) to deal effectively with those problems.[95] Environmental changes are a prominent example of the types of problems Brown describes. Brown has observed that among the responses to this incongruity is a greater degree of accountability across national borders than has existed in the past: "The normative core of such a globalized accountability system would be the principle that *those who can or do substantially affect the security or well-being of others (especially inflicting harm) are assumed to be accountable to those they can or do affect.*"[96] Agreements signed at the Earth Summit are a manifestation of this nascent accountability. They set in motion a process in which the North takes greater responsibility for the harm done to the entire world as a consequence of industrial development and affluent lifestyles, and they go one step further in recognizing the need for the North to aid the South in achieving sustainable development. As Brown points out, while the Bush administration gave only grudging support to the UNCED initiatives, the Clinton administration did more by joining its Group of Seven partners in modest funding of the Global Environment Facility and by agreeing to restructuring it to give the developing countries somewhat more influence when decisions regarding the allocation of funds are made.[97]

More recent examples of apparently altruistic concerns of the people and governments of developed countries were the "Jubilee 2000" campaign to cancel the foreign debt of the world's poorest countries, the associated 1999 declaration by G-7 countries to eliminate $100 billion in poor country debt, increasing efforts by the United States and international financial institutions to mitigate the adverse impacts for the world's poor of economic globalization, and indications that "equity has assumed an equal place alongside economic growth and stability" in the IMF, World Bank, UNDP, and other international development agencies.[98] Indeed, in the closing weeks of his tenure, President Clinton spoke of "responsibility in the developed world to shape this process [of globalization] so that it lifts people in all nations."[99]

Citing altruism as a key determinant of international aid from Western governments, one Northern diplomat who participated in the UNCED deliberations wrote:

> One could tentatively hypothesize that something in the nature of the environmental subject matter (such as precisely the feeling that one is contributing to the "common good" and "the welfare of future generations") tends to make politicians and negotiators readier to look for common ground than they would be in other sorts of international negotiation. This, of course, certainly does not mean that they will be willing to abandon their national interest, simply that they may be willing to bend it slightly more than they would in other circumstances to get an agreed outcome. Certainly something of that feeling emerges from the records of the chief U.S. negotiator in the ozone layer negotiation [Richard E. Benedick], and my personal experience of various EC negotiations as well as the climate change convention has given me something of the same impression.[100]

Shortly after the Earth Summit, EPA director William Reilly wrote that in his opinion, "a transition is taking place: countries with enormous economic resources are beginning to acknowledge social and environmental obligations commensurate with their economic power."[101] As noted in the first chapter, members of the Clinton administration also began to acknowledge U.S. accountability in this issue area. For example, then Department of State counselor Timothy Wirth said that the Clinton administration would place the United States "out in the lead" in "fulfilling its responsibilities" in the follow-up to UNCED.[102] Vice President Gore said that the United States had a "responsibility to deal with [its] disproportionate impact" on the global environment.[103] And President Clinton acknowledged that Americans had "common responsibilities" with regard to sustainable development and the growing global gap between rich and poor.[104]

Conclusion

Consideration of international equity as an instrument to promote national interests or to fulfill demands of politically influential actors are not the only explanations for why the United States (and many other developed countries) supported in varying degrees the equity considerations that were part of UNCED and other international environmental deliberations. Governments, leaders, decision makers, diplomats, and bureaucrats may act to promote the interests of others; they may from time to time act altruistically. One objective of governments and individual policymakers may be to do good, to help others, and to do what is fair and just. Equity can be viewed as a value in itself.

In the case of individuals active in the foreign policy process, efforts to introduce or support considerations of equity in international environmental deliberations may coincide with or run contrary to perceived or stated national interests. Alternatively, their government may have no particular policy, leaving the decision of whether to promote equity—and how much—to individuals in the policy-making process. Government officials may be concerned about their own and their government's reputations from the perspectives of other governments and policy elites, as well as from the perspective of the scholarly community. In the United States especially, policymakers have historically felt compelled to justify actions of U.S. governments in idealistic terms, even when those actions are taken for largely or completely self-interested reasons. As it has become less acceptable to treat developing countries inequitably, at least in the context of environmental change, the U.S. government and particularly individuals in it have been (and may be in the future) pushed in subtle ways toward policies that accept international equity as an objective of U.S. policy. Insofar as other developed country governments act in accordance with nascent norms of international equity, and insofar as policy elites and academics in the United States advocate that the U.S. government adopt a similar posture for reasons of national interest and ethics, it seems likely that the U.S. government will increasingly accept international equity as one objective of its foreign policy, in the short run especially in the environmental policy field. In the case of UNCED and subsequent international environmental deliberations, increasing international concern for international equity has been the unavoidable context in which the United States has operated.

Equity may also be a long-term result of otherwise self-interested actions. That is, as governments and policymakers increasingly use equity as a means to promote their self-interests, it can become institutionalized with a driving force of its own. International equity can become an operative norm and a preferred way for states to "do business." Relative to its use as direct or indirect means to self-interested ends, the altruistic promotion of international equity considerations still takes a back seat. Nevertheless, equity as a value in itself is a variable that can operate as an independent force. For example, for decades the Scandinavian governments and diplomats have called for steps to promote international equity. Such calls have been echoed in their statements related to UNCED and other international environmental deliberations. Most important, they have "put their money where their mouths are" by transferring on the order of one percent of their GNPs to poor countries in ways clearly meant first and foremost to promote the well-being of poor persons there. While the United States has been much less generous, even in the environmental field, it has begun to move in the direction of promoting welfare in the poor countries, at least insofar as doing so can contribute to sustainable development. Only the most cynical and uninformed observer would attribute *all* U.S. foreign aid to self-interest. A significant—usually small but sometimes substantial—portion is attributable to the simple desire to do the right thing.

Compared to the previous two chapters, this chapter has suggested more subtle explanations for equity considerations in the UNCED process: the industrialized democracies—including the United States—are, in short, rather generous with their money, at least compared to their beneficence in centuries past—and even compared to the first half of the twentieth century. (They still tend to be more generous with their pledges than with their actual spending, of course.) Individuals within the policy process are sometimes interested in promoting international equity *per se*. Economic justice and beneficence are universal ethics that are sometimes promoted by the countries with the resources to do so.[105] If we look back at the evolution of equity in international environmental deliberations, we see a process similar to the evolution of the foreign aid regime. Thus, "Although at Stockholm only lip service was given to the notion that economic development was the solution to environmental degradation in the Third World," according to Haas et al., "this idea was the bedrock of the Rio conference both in name and as enunciated in principle 3 of the Rio Declaration (despite U.S. resistance) asserting 'the right to development.'"[106]

To be sure, the United States could have done much more. But it could have done even less than it did. It could have taken less responsibility for global pollution, refused to agree to provisions for "new and additional" funds and transfers of technology on concessional terms, and it could have vetoed the right of developing countries to have somewhat more say in who gets what funds for sustainable development. Most important, the Clinton administration gradually moved the United States in the direction of doing substantially more to promote international equity in the context of sustainable development, despite the administration's very limited resources when facing a hostile Congress. The upshot is that the U.S. government started to accept international equity as an objective of global environmental policy, in part because those involved in the policy process were sometimes sympathetic to the altruistic objectives of the foreign aid regime and because they viewed international environmental equity as one reasonable means by which to promote the general welfare.

Notes

1   Arnold P. Shifferdecker, former director of the Office of Environmental Protection, Bureau of Oceans and International Environmental and Scientific Affairs, U.S. Department of State, letter to the author, n.d. (received 21 April 1992).

2   Philip Shabecoff, *A New Name for Peace: International Environmentalism, Sustainable Development, and Democracy* (Hanover, NH: University Press of New England, 1996), pp. 135-36.

3   Susan F. Drake, former advisor to U.S. UNCED delegation and officer in the Office of Environmental Protection, Bureau of Oceans and International Environmental and Scientific Affairs, Department of State, telephone interview by author, 1 November 1995. Drake named several individuals in the U.S. government sympathetic to equity considerations in the context of UNCED, including Robert Ryan, Melinda Chandler, and E.U. Curtis "Buff" Bohlen (and one other whose name she could not recall at the time of the interview).

4   David H. Lumsdaine, *Moral Vision in International Politics: The Foreign Aid Regime, 1949-1989* (Princeton: Princeton University Press, 1993) p. 10.

5   U.S., Congress, House, Committee on Foreign Affairs, Subcommittee on Human Rights and International Organizations, testimony of Elliot L. Richardson, Co-Chairman, National Council, U.N. Association of the U.S.A., *U.S. Policy Toward the 1992 United Nations Conference on Environment and Development*, 102nd Cong. [17 April, 24 July, and 3 October 1991], 1992, pp. 96-97. In the same hearing, former Secretary of State Edmund Muskie said that on the U.S.'s UNCED preparations, "it is fairly clear to me that the International Organizations Desk in the State Department basically is equal to the Environmental Desk and they neutralize each other. They just take different positions and cannot come up with a coordinated position" (p. 101).

6   Not for attribution interview by author with U.S. UNCED delegate, 12 January 1996.

7   Drake.

8   Scott A. Hajost, "The Role of the United States" in Luigi Campiglio et al., eds., *The Environment After Rio: International Law and Economics* (London: Graham and Trotman, 1994), p. 17. Note how Hajost seems to express the view that at least some of the developing countries' concerns were "legitimate." This parallels Susan Drake's views on developing country concerns described below. William Reilly, EPA administrator and head of the U.S. delegation to Rio, sent a long memo to EPA employees after the summit. He wrote that the U.S. government "assigned a low priority to the negotiations of the biodiversity treaty, [was] slow to engage the climate issue, [was] last to commit our President to attend Rio. We put our delegation together late and we committed few resources." William K. Reilly, "Reflections on the Earth Summit," memorandum to all EPA employees, 15 July 1992, reprinted in U.S., Congress, House, Joint Hearings, Committee on Foreign Affairs and Committee on Merchant Marines and Fisheries, *U.S. Policy Toward the United Nations Conference*

on Environment and Development, 102nd Cong., 2nd. sess., 26, 27 February, 21, 28 July 1992, p. 399.

9   Drake.

10   Ibid.

11   Ibid. Drake said that the other participants in the meeting were diplomats from Mexico, Tunisia (representing the G-77), India, France (the EC Representative), Canada, Sweden, and Norway. For a description of the international equity provisions of Resolution 44/228, see the section in chapter 4 titled "Preparations for the Earth Summit" and United Nations General Assembly Resolution 44/228, "United Nations Conference on Environment and Development," A/44/228, 22 December 1989. The secret meeting described by Drake became public and led to angry calls for "transparency" at the United Nations. United States UN Representative Thomas Pickering and his deputy John Moore knew about the meeting, according to Drake, but they refused to admit their knowledge and she "had to take the heat." Regarding the differences of opinion in the White House and State Department over equity considerations, Drake said that "personalities make up 90 percent of this stuff." Drake's account of Sununu's power in the bureaucracy is supported by Gore, who has written that President Bush allowed Sununu "not only to set [global environmental] policy in Bush's name but also to stifle disagreement within the administration." Al Gore, *Earth in the Balance* (New York: Houghton Mifflin, 1992), p. 175.

12   Drake.

13   Ibid.

14   Quoted in Pratap Chatterjee and Mathias Finger, *The Earth Brokers: Power, Politics and World Development* (London: Routledge, 1994), p. 138. Drake (interview) identifies Bohlen as one of those persons in the Bush administration that were sympathetic to considerations of North-South equity.

15   Gunnar Sjostedt, "Issue Clarification and the Role of Consensual Knowledge in the UNCED Process" in Bertram I. Spector, Gunnar Sjostedt and I. William Zartman, eds., *Negotiating International Regimes: Lessons Learned from UNCED* (London: Graham and Trotman, 1994), p. 70.

16   Stanley P. Johnson, *The Earth Summit* (London: Graham and Trotman, 1993), p. 11. Maurice Strong, Secretary General of UNCED, reminded delegates to the first prepcom that "the key sectoral issues of financial resources, technology transfer and institutions must have the especially important place on the Conference agenda provided for in General Assembly Resolution 44/228." "Report of the Secretary General of the Conference," A/Conf.151/PC/5/Add.1, 6 August 1990.

17   Chris K. Mensah [Member of the G-77 UNCED Negotiating Team on Legal Instruments and Member of Vanuatu Delegation to UNCED], "The Role of Developing Countries" in Luigi Campiglio, et al., eds., *The Environment After Rio: International Law and Economics* (London: Graham and Trotman, 1994), p. 37. See also South Centre on Environment and Development, *Towards a Common Strategy for the South in the UNCED Negotiations and Beyond*, Geneva, November 1991.

18   Drake. Drake's assertion is backed up by the Kuala Lumpur Declaration of developing countries at which ministers from developing countries reaffirmed their determination to see that the Earth Summit "fully and clearly achieves in concrete terms, the provisions contained in the relevant General Assembly resolutions, notably, UN/GA Resolution 44/228." See Second Ministerial Conference of Developing Countries on Environment and Development, "UNCED: Kuala Lumpur Declaration on Environment and Development," 29 April 1992. The most important and controversial

chapter of Agenda 21 (Chapter 33) quotes extensively from the financial provisions of 44/228 (regarding new and additional funds, special regard for needs of developing countries, and transfer of technology). See Agenda 21, para. 33.1.

19   Johnson, p. 448.
20   Robert Ryan, Jr., former Director, U.S. Office of the United Nations Conference on Environment and Development, Department of State, telephone interview by author, 12 January 1996.
21   Ibid. For example, the U.S. UNCED coordinator attended a meeting at the Woodstock Theological Center dealing with ethical issues surrounding UNCED.
22   Ibid.
23   U.S., *U.S. Policy Toward the 1992 United Nations Conference on Environment and Development*, cited in statement of Gareth Porter, Environmental and Energy Study Institute, 16 September 1991, p. 266.
24   Shabecoff, p. 137. According to Shabecoff, most members of the U.S. working UNCED delegation, led by ambassadors E.U. Curtis Bohlen and Robert Ryan, were highly respected by other diplomats, and were not held responsible for the positions they brought to the negotiations. Indeed, at the fourth prepcom Bohlen delivered a speech in which he committed the U.S. to the principle of new and additional financial resources to help the poor countries develop sustainably. It turned out, however, that Bohlen did not have support for such a change in U.S. position from the White House. Ibid., pp. 154-55.
25   According to Robert Ryan, "lots of people in the Clinton administration are personally sympathetic toward helping poor people." Ryan.
26   As far as the author can determine, Gore was the only member of either house of Congress to participate in all UNCED prepcoms.
27   Gore, *Earth in the Balance*, pp. 220, 295-360. There was a remarkable similarity between the "Global Marshall Plan" described in Gore's book and the environmental foreign policy of the U.S. government under President Clinton. In short, the plan called for stabilizing world population, developing and sharing environmentally appropriate technologies, sustainable development (integration of environmental considerations into economic practices), a new generation of environmental treaties and agreements (e.g., Montreal Protocol, FCCC, Biodiversity Convention, etc.), and a global education program to spread understanding of the global environment. Ibid., pp. 305-60.
28   Ibid., p. 269.
29   Gore immediately continued: "Without such commitments [to social justice], the world cannot contemplate the all out effort urgently needed." Ibid, p. 279.
30   Ibid., p. 297.
31   Ibid., p. 300.
32   Ibid., p. 96. He added: "I favor a international treaty limiting the amounts of CO_2 individual nations are entitled to produce each year. . ." (p. 345).
33   Clinton had a poor record of environmental protection in Arkansas, most notably his involvement in chicken farms that severely polluted nearby waterways.
34   John M. Broder and Melinda Henneberger, "Few in No. 2 Spot Have Been as Involved as Gore," *New York Times*, 31 October 2000.
35   Ibid.
36   Robert W. McElroy, *Morality and American Foreign Policy* (Princeton: Princeton University Press, 1992), pp. 42-43.

37 David L. Cingranelli, *Ethics, American Foreign Policy, and the Third World* (New York: St. Martin's Press, 1993), p. 21.

38 Ibid.

39 "Highlights from a Review of Existing Survey Data Regarding American Views on U.S. Leadership and Foreign Assistance, Summary Findings," *Polls and Public Opinion: The Myth of Opposition to Foreign Assistance*, May 1994, <gopher://gaia.infousaid.gov/0r0-13682-/agancy_wide/why_foreign_aid/pol>. Other data seem to indicate a decline in support for humanitarian goals. In a 1994 poll, 56 percent of the public and 41 percent of the "leaders" polled thought that combating world hunger should be a very important foreign policy goal of the U.S. Only 22 percent of the public and 28 percent of the leaders thought the same of "helping to improve the standard of living of less developed countries." See John E. Reilly, "The Public Mood at Mid-Decade," *Foreign Policy* (Spring 1995), p. 82. Support in 1993 for using U.S. troops to prevent famine and starvation may have been influenced by U.S. intervention in Somalia in December 1992. Indeed, President Bush called the U.S. intervention "God's work." According to the U.S. Agency for International Development, foreign aid is in the U.S.'s "own interest. It contributes to the growth of our economy. Americans continue to have a humanitarian desire to help the less fortunate. We must address problems that cross borders such as the environment, narcotics, and AIDS. We have an interest in a peaceful, stable world." U.S. Agency for International Development, *Why Foreign Aid? The Benefit of Foreign Assistance to the United States* (Washington: AID, February 1992).

40 This opinion was held strongly by three in ten Americans (31 percent). "Purpose of Foreign Policy: National Interest or Human Values?," from The Gallup Opinion Monitor, *National Security* 1, 1 (July 1993).

41 This attitude crossed party lines: 78 percent of Republican respondents agreed. Program on International Policy Attitudes, Joint Program of the Center for Study of Policy Attitudes and the Center for International and Security Studies of the University of Maryland, *Americans and Foreign Aid*, 23 January 1995.

42 Chicago Council on Foreign Relations, "American Public Opinion and U.S. Foreign Policy 1999," March 1999, <http://www.ccfr.org/publications/opinion/opinion.html>.

43 Ian Smillie, "Mixed Messages: Public Opinion and Development Assistance," paper delivered to the Organization for Economic Cooperation and Development, 25 October 1994.

44 Steven Kull and Clay Ramsay, "Challenging U.S. Policymakers' Image of an Isolationist Public," *International Studies Perspectives* 1 (2000), p. 110 (my emphasis).

45 Ibid., pp. 110-11.

46 Lumsdaine, p. 287.

47 Ibid., p. 36.

48 See, for example, Lumsdaine; Alain Noel and Jean-Philippe Therien, "From Domestic to International Justice: The Welfare State and Foreign Aid," *International Organization* 49, 3 (Summer 1995), pp. 523-53; and Louis-Marie Imbeau, *Donor Aid- The Determinants of Development Allocations to Third World Countries: A Comparative Analysis* (New York: Peter Lang, 1989).

49 Noel and Therien, p. 523. (United Nations Development Program, *Human Development Report 1992* [New York: Oxford University Press, 1992] is cited as the original source.)

50 Ibid., p. 524.

51 Gunnar Myrdal, *Beyond the Welfare State* (New Haven: Yale University Press, 1960).

52 Gunnar Myrdal, *The Challenge of World Poverty* (New York: Vintage Books, 1970), p. 360, original emphasis.

53 Myrdal, *The Challenge of World Poverty*, p. 275.

54 Lumsdaine, p. 33.

55 Ibid., p. 290.

56 Payaslian found that in the Reagan and Bush administrations, humanitarian concerns were important determinants for development assistance, but less so for military aid. Simon Payaslian, *U.S. Foreign Economic and Military Aid: The Reagan and Bush Administrations* (Lanham: University Press of America, 1996). Lumsdaine argues that about half of U.S. foreign assistance was motivated by security concerns and other national interests. This helps explain why recently U.S. foreign aid has dropped more sharply than that of most other developed countries. With the end of the cold war, the rationale for much U.S. aid disappeared, whereas the rationale for much aid from most other OECD countries—in part to help those in need, in part to promote trade—remained unchanged.

57 Lumsdaine, p. 3.

58 Ibid., p. 6. Lumsdaine's conclusions are controversial (they are backed, however, by substantial statistical analysis). They are important here because they show that beneficence was an important component of OECD foreign aid policies generally. Although this was less true of the U.S. case, the United States was forced to operate in a context in which its allies were directing much of their foreign assistance to programs meant to promote the general welfare, as opposed to focusing all aid on programs that would redound to the donors.

59 Ibid., p. 5.

60 Ibid., p. 268.

61 Ibid., p. 37.

62 Ibid., p. 47.

63 Very poor countries are defined as those with per capita incomes of less than about $500. David W. Pearce, ed., *The M.I.T. Dictionary of Modern Economics* (Cambridge, MA: MIT Press, 1992), p. 213. (Replenishment of IDA funds, including an additional "earth increment" of $5 billion [$1.5 billion from IBRD interest, $3.5 billion from donors], was discussed in UNCED negotiations.)

64 Lumsdaine, pp. 40-41; U.S. aid to multilateral institutions was even higher (31 percent) in the early 1980s. See Lumsdaine, Table 2.4.

65 Ibid., p. 48.

66 Ibid., p. 66.

67 Ibid., p. 67.

68 United Nations Development Program (UNDP), *Human Development Report 1994* (Oxford: Oxford University Press, 1994), p. 71.

69 See Gosta Esping-Anderson, *The Three Worlds of Welfare Capitalism* (Princeton: Princeton University Press, 1990).

70 Noel and Therien, pp. 538-39.

71 Ibid., p. 539.

72 Ibid.

73 Ibid., p. 551.

74 See the special year-end review issue of *The Earth Times* (25 December 1995-14 January 1996) containing numerous articles on the Danish model. Several contributions are cited below.

75   In 1993 both Sweden and Denmark's development aid exceeded one percent of their GNPs. C. Gerald Fraser, "The Mechanics: How to Organize Giving," *The Earth Times* (25 December 1995-14 January 1996), p. 13.

76   Daniel J. Shepard, "Denmark 1995: Setting a Global Standard," *The Earth Times* (25 December 1995-14 January 1996), p. 11.

77   Half of Denmark's bilateral assistance goes to about 20 poor countries, including Tanzania, Uganda, India, Mozambique, Zimbabwe, Nepal, Egypt, Kenya, and Nicaragua. Fraser, p. 13.

78   Pernille Tranberg, "No Worry About Security," *The Earth Times* (25 December 1995-14 January 1996), p. 12.

79   Pernille Tranberg, "It's a Cultural Thing: A Willingness to Pay," *The Earth Times* (25 December 1995-14 January 1996), p. 13.

80   Ryan. Increasingly, other developed countries are taking the lead in funding for the world's poor. In the mid-1990s, the EU gave about three times as much in development aid as the United States, despite having about the same size population.

81   To put this into perspective, world military spending was $815 billion in 1992. United Nations Development Program (UNDP), *Human Development Report 1994* (Oxford: Oxford University Press, 1994), pp. 62, 48.

82   In ball-park figures, the amount of money estimated necessary to implement Agenda 21 sustainable development programs in the developing counties is $600 billion, with $125 billion of that expected to come from the developed countries. See Chapter 33, para. 33.18, of Agenda 21 in United Nations, *Report of the United Nations Conference on Environment and Development, Volume I: Resolutions Adopted by the Conference* (New York: United Nations, 1993), pp. 9-479. Not surprisingly, therefore, UNCED Secretary General Maurice Strong suggested that ODA be doubled to 0.7 percent of developed countries' GNPs, thus approaching the $125 billion estimate.

83   Noel and Therien, pp. 534-37. Average net ODA disbursed in 1982-83 was 0.23 percent; in 1993 it was 0.30 percent. UNDP, *Human Development Report* (New York: Oxford University Press, 1995), table 29, p. 204. All this suggests, in short, that ODA waxes and wanes, hovering near one-third of one percent of the developed countries GNPs. This seems to justify directing our attention to the *objectives* of that aid. Increasingly, the objective is to meet the basic needs of the world's poor.

84   This ranged among OECD countries in 1990 from .16 percent in the case of Ireland to 1.17 percent in the case of Norway. "Rio Conference on Environment and Development," *Environmental Policy and Law* 22, 4 (1992), p. 218.

85   World Bank, *Development Brief* 16 (April 1993), p. 1; OECD, "Sharp Changes in the Structure of Financial Flows to Developing Countries and Countries in Transition," *OECD Press Release*, 24 June 1994.

86   Shabecoff, *A New Name for Peace*, p. 186-87.

87   Congressional Research Service, *International Environment: Briefing Book on Major Selected Issues*, report prepared for the House Committee on Foreign Affairs, July 1993, p. 88. For a discussion of the financing question before the Rio summit, see Congressional Research Service, *Financing New International Environmental Commitments*, report prepared for the House Committee on Foreign Affairs and the Senate Committee on Foreign Relations, March 1992.

88   Lumsdaine, p. 36. Private investment has increased rapidly in recent years, reaching $173 billion in 1994, four times what it was in 1989, and more than four times the volume of World Bank lending in 1994. World Bank statistics cited in Hilary French,

Partnership for the Planet, Worldwatch Paper No. 126 (Washington: Worldwatch Institute, July 1995), p. 39.

89 Cf. Noel and Therien, p. 528; Lumsdaine, passim.

90 The 20/20 concept was adopted as a nonbinding goal only. See, for example, Eva Jespersen, "Why 20/20?" paper presented at Harvard University Center for Population and Development Studies, 28 June 1994. The World Summit for Social Development, held in March 1995, is described in "A Final Report on the World Summit for Social Development," *Earth Negotiations Bulletin* 10, 45 (18 December 1995).

91 Daniel J. Shepard, "Assessing the Results of the Social Summit," *The Earth Summit Times* (25 December 1995-14 January 1996), p. 12.

92 Michael Grubb et al., *The Earth Summit Agreements: A Guide and Assessment* (London: Earthscan, 1993), p. 33.

93 See the chapters by Ahmed Djoghlaf, Chandrashekhar Dasgupta, Bo Kjellen, and Nitze in Irving M. Mintzner and J.A. Leonard, *Negotiating Climate Change: The Inside Story of the Rio Convention* (Cambridge, England: Cambridge University Press, 1994).

94 An exception was not made in 1993. In that year the U.S. spent 0.15 percent of its GNP on ODA; Ireland spent 0.20 percent of its GNP for ODA. UNDP, *Human Development Report 1995*, p. 204.

95 Seyom Brown, *New Forces, Old Forces, and the Future of World Politics* (New York: Harper Collins, 1995). The incongruence described by Brown is a central theme in James Rosenau, *Turbulence in World Politics* (Princeton: Princeton University Press, 1990).

96 Brown, p. 255, original emphasis.

97 Ibid., pp. 264-65.

98 Steven Pearlstein, "In Prague, Capitalism with a Human Face," *International Herald Tribune*, 2 October 2000.

99 Marc Lacey, "Clinton Reminds Leaders of Their Duty to the Poor," *New York Times*, 14 December 2000.

100 Tony Brenton, *The Greening of Machiavelli* (London: Earthscan Publications, 1994), pp. 258-59. Brenton was head of the UK Foreign and Commonwealth Office bureau dealing with international environmental legislation in the run-up to the Earth Summit.

101 William K. Reilly, "Reflections on the Earth Summit," memorandum to all EPA employees, 15 July 1992, reprinted in U.S., *U.S. Policy Toward the United Nations Conference on Environment and Development*, p. 399.

102 Timothy Wirth, "World Conference on Human Rights," press briefing, Washington, DC, 2 June 1993, U.S. Department of State, *Dispatch* 4, 23 (7 June 1993).

103 Albert Gore, "U.S. Support for Global Commitment to Sustainable Development," address to the Commission on Sustainable Development, United Nations, New York City, 14 June 1993, *U.S. Department of State Dispatch* 4, 24 (14 June 1993).

104 William J. Clinton, "Advancing a Vision of Sustainable Development," address to the National Academy of Sciences, Washington, DC, 29 June 1994, U.S. Department of State, *Dispatch* 5, 29 (18 July 1994).

105 A.J.M. Milne, *Human Rights and Human Diversity* (Albany: State University of New York Press, 1986); Frances V. Harbour, "Basic Moral Values: A Shared Core," *Ethics and International Affairs* 9 (1995), pp. 155-70; Donald J. Pachula, "The Ethics of Globalism," *ACUNS [Academic Council on the United Nations System] Reports and Papers* 3 (1995), pp. 1-22.

106 Peter M. Haas, Marc A. Levy and Edward A. Parson, "Appraising the Earth Summit: How Should We Judge UNCED's Success?" in Ken Conca et al., eds., *Green Planet Blues: Environmental Politics from Stockholm to Rio* (Boulder: Westview Press, 1995), p. 163.

PART III

EQUITY, U.S. FOREIGN POLICY, AND THE FUTURE OF GLOBAL ENVIRONMENTAL POLITICS

9 International Equity and Global Environmental Change: Implications for American Foreign Policy and Humankind

For much of the last century, the United States was frequently able to influence international affairs to its advantage by using traditional power capabilities: its strong military and the large, usually vibrant U.S. economy that supports it. However, in a globalized world, such capabilities are much less fungible; it is sometimes—or usually—difficult for the United States to translate its traditional capabilities into concrete influence. International efforts to address changes to the global environment demonstrate this changing situation. The United States, despite its superior economy and vast military arsenal, cannot force other countries to protect environments and global commons (e.g., the atmosphere and oceans) in which the United States, like most countries, has interests. The same can be said for other countries that have "great power" status.

The ongoing climate change conferences, such as those at Kyoto in 1997, Buenos Aires in 1998, and The Hague in 2000, demonstrate the limitations of great power capabilities in the environmental issue area. The United States, the world's sole "superpower," could not and cannot resist taking on a fairer share of the climate change burdens, nor could it tell the economically developing countries "reduce your emissions of greenhouse gases—or else." Nuclear bombs, aircraft carriers, or other military capabilities, to use the extreme examples, are simply of no use in this situation.[1] Neither could the United States completely ignore other countries' demands or use draconian measures such as threats of economic sanctions or trade wars to get its way. It had to compromise with "weaker" countries. These events suggest that the United States must find more subtle methods for influencing the behavior of other countries.

This chapter argues that the United States can more effectively achieve its foreign policy goals in the environmental issue area, and thereby protect and promote U.S. national interests, by treating other countries, particularly the developing countries, more equitably.[2] The United States has yet to fully comprehend the changes in power capabilities inherent to contemporary international affairs. Yet, given the will and careful deployment, its financial, technological, and diplomatic resources can be brought to bear on problems of environment and development. Such policies would also promote ethical imperatives, such as taking responsibility for America's disproportionate adverse impacts on the global environment. The United States is the world's greatest polluter; as such, it has the greatest ethical obligation to redress the wrongs of historical and ongoing global pollution. Thus, with regard to environmentally sustainable development, the United States can further its national interests, and, coincidentally, promote ethical goals, by meeting many of the demands from developing countries (and more and more frequently from developed countries) for greater international environmental equity, defined here as a fair and just distribution of the benefits, burdens, and decision making authority associated with international environmental issues.

Environmental Change and International Equity

Effective efforts to implement sustainable economic development on a global basis require that demands from the developing countries for equity receive serious consideration by the United States and other industrialized countries, and indeed that equity be implemented as part of global environmental protection measures.[3] Recent international environmental agreements have codified notions of equity, suggesting what "equity" could look like in reality. Examples include requirements for new and additional funds to assist developing countries in meeting the provisions of international environmental agreements, new or changed international environmental funding institutions that give developing countries greater authority in decisions regarding the allocation of funds for sustainable development (e.g., the ozone multilateral fund and the Global Environment Facility), and requirements of the ozone and climate change regimes for developed countries to act before, and more forcefully than, developing countries to address these problems.

    In the context of global environmental change, the United States and other industrialized countries have started to give international equity and fairness—and particularly the demands by economically less developed countries for greater international equity—more consideration.[4] This change is clearer when contrasted with the developing countries' unsuccessful calls in the 1960s and 1970s for greater equity via a New International Economic Order. The United States is among those countries that have come to realize that the developing world cannot usually be coerced into cooperating on matters of sustainable development and global environmental protection. Rather, developing countries must be persuaded to cooperate, in part by meeting some of their demands for more equitable treatment in international environmental affairs. Toward this end, the United States has softened its historic opposition to concessional financial and technological assistance to developing countries—although too much opposition remains. To be sure, like most other developed countries, the United States has cut its foreign assistance spending.[5] Nevertheless, funds have often been redistributed to allow increases in funding for sustainable development programs.

Concerning climate change specifically, in 1998 the Clinton administration signed the Kyoto Protocol to the Framework Convention on Climate Change (FCCC). During the 1997 negotiations that produced the Protocol, U.S. diplomats insisted that developing countries take on "new" and "meaningful" obligations, an arguably inequitable stance. Nevertheless, in the event the United States agreed to an instrument that—at the insistence of traditionally weak countries—made no mention of such obligations. Instead, the treaty requires *developed* countries to reduce their greenhouse gas emissions by about five percent in aggregate. The United States and other developed countries appropriately agreed to take on all the initial burdens of addressing climate change.

Some great powers have realized that being seen as taking on their fair share of global environmental burdens can have benefits. The West Europeans viewed the 1992 Earth Summit as an avenue to exercise new influence toward the developing world. They capitalized on the U.S. government's failure to take a leadership role at the Earth Summit. More recently, the Labour government of Tony Blair in Britain has sought to lead the industrialized world in the climate change negotiations. It insists that it will reduce its greenhouse gases largely in line with the demands of many developing countries, thereby showing its commitment to combat climate change before demanding that the poorer countries take on substantial emissions limitations of their own. During the sixth conference of the parties to the Climate Change Convention in 2000, the European Union countries joined developing countries in resisting efforts by the United States to ease its burdens under the Kyoto Protocol. Of course, some developed countries have acted quite differently: using its own convoluted case for equity relative to other developed countries, Australia successfully argued during the Kyoto negotiations that it should be allowed to *increase* its greenhouse gas emissions.

The U.S. government—particularly the Clinton administration—gradually came to accept equity as an important objective of its global environmental policy. However, the degree to which it implemented this idea was limited and no doubt is a great disappointment to environmentalists and advocates of development assistance. In short, as the previous three chapters argued, recent American willingness to accept the importance of equity in international environmental politics—or at least to not actively and effectively prevent its incorporation into international environmental instruments (including the climate change agreements), as happened more systematically before the advent of Clinton—is a function of many factors: concern for U.S. economic and environmental security, domestic and international political pressures, and the appeal of the notion of equity as a value in itself.

To be sure, the United States government has not gone as far as developing countries want—and indeed as far as many other developed countries, particularly EU member states, apparently want. But recent U.S. policy edges toward the emerging consensus among both developed and developing countries that equity is an important part of international environmental protection efforts. While we are far from seeing the developing countries demands for a New International Economic Order realized, adverse environmental changes, ranging from local problems like water pollution and deforestation to (especially) global problems like ozone depletion and climate change, have nevertheless acted as stimuli for considerations of equity in international relations and U.S. foreign policy. Because the developing countries are essential to effective international environmental protection that affects the United States and indeed most countries, notions like equity are increasingly seen as useful instruments for the creation and effectiveness of international environmental institutions.[6] This is unusual by historical standards.

Ongoing deliberations regarding global environmental issues will probably be among the most important forums for serious discussions of international equity. To be sure, other international conferences have tackled considerations of international equity (e.g., the United Nations World Summit for Social Development held in Copenhagen in 1995). But it may be in the environmental issue area that the most progress is made, in large measure because so many countries may suffer the consequences of global environmental change, and because the developed countries—which are best equipped to take effective action within their own jurisdictions and to finance action in less affluent countries—have recognized the degree to which poverty and traditional industrialization in the poor countries may affect their own environmental security and economic vitality. Environmental issues, more than any others, are compelling governments and diplomats the world over to seriously consider equity.[7] In short, environmental changes have raised the salience and profile of equity in international relations.

This is not to say that the developing countries' battle has been won—far from it. Rhetoric has clearly outpaced tangible policy. Nevertheless, considerations of equity have started to take on a life of their own. Even outside the realm of international environmental affairs, it is no longer easy for the United States (or other great powers) to treat weaker and poorer countries in ways that contradict commonly accepted notions of fairness. To do so could result in unfavorable responses not only from the victims, who—especially with regard to environmental matters—have the capacity to hurt the United States in the long-term, but also from other developed countries, thereby adversely affecting U.S. influence in international politics generally. Examples abound in the international environmental area: the Montreal Protocol on Substances that Deplete the Ozone Layer contains equity provisions (e.g., grace periods for developing countries and a multilateral development fund), and the FCCC and related agreements provide for "common but differentiated responsibility" (meaning the developed countries must act first and aid the developing countries if they are to take action),[8] in order to garner willing participation of the developing countries in these instruments.

With the foregoing in mind, the remainder of this chapter examines some of the practical and normative implications of adopting and embracing equity as an objective of global environmental policy—and particularly policy on global climate change—for the United States and for the world, as well as the implications that such a policy would have for American global power in this century. Many of the general examples here, and certainly the conclusions, apply to other traditional great powers— particularly countries that hope to remain or become "great" in the future.

International Environmental Equity and U.S. Foreign Policy: Some Practical Implications

By promoting considerations of international environmental equity the United States can help protect its national interests adversely affected by environmental change and related poverty in the developing world.[9] The Clinton administration tried to move the United States in this direction, at least relative to historical standards of U.S. foreign policy. But substantial financial and diplomatic resources did not follow the administration's rhetoric. It failed to muster the robust support of the entire foreign policy establishment, and it faced tremendous hurdles, most notably in convincing the Republican-controlled Congress of the merits of funding this type of policy. The Congress has been generally opposed to new foreign aid; foreign assistance spending continues to fall in the United States, already dipping below .08 percent of GDP—compared to over one percent in some West European and Nordic countries.[10]

In failing to embrace international equity, the United States threatens to develop a bad reputation in the eyes of many other international actors, logically leading to questions about the extent to which the United States can be trusted to treat them fairly and equitably in the cooperative arrangements that are increasingly important to promoting U.S. goals in international affairs. By becoming a leader with regard to considerations of international environmental equity and international equity generally, the United States could bolster its global influence and power in the coming decades.

Environmental Changes Threaten National Interests

While the U.S. government has slowly come to recognize that environmental changes can threaten American interests, it has failed to give those threats the high priority (relative to traditional threats) that they require, and it has been slow to recognize that by promoting international environmental equity it can reduce the likelihood that those threats will adversely affect U.S. interests.

Increasingly, the military threats to U.S. interests that characterized the cold war are being joined or replaced by threats from environmental change and related economic changes impacting on the U.S. economy, the American way of life, and overseas interests.[11] Climate change, loss of species, ozone depletion, ocean pollution and the like pose direct threats to the health and well-being of Americans. These and other more local environmental changes can also reduce the ability of other countries, particularly developing countries, to purchase American exports, thus affecting the U.S. economy. Former Secretary of State Warren Christopher pointed out that by the end of the century 80 percent of the world's consumers will live in the developing countries. He said that the United States ought to give greater attention to the "interlocking threats of insupportable population growth, endemic poverty and environmental degradation."[12] If the United States fails to do so, according to Christopher, "the result will be widespread suffering abroad and the loss of export opportunities for American companies, worker and farmers."[13] Secretary Christopher's successor, Madeleine Albright, echoed these comments when she said that "damage to the global environment, whether it is over fishing of the oceans, the build-up of greenhouse gases in the atmosphere, the release of chemical pollutants, or the destruction of tropical forests, threatens the health of the American people and the future of our economy. [E]nvironmental problems are often at the heart of the political and economic challenges we face around the world."[14]

Environmental changes can contribute to migrations of large numbers of individuals to the United States and other developed countries, as well as to nearby developing countries even less able to deal with their arrival, contributing to domestic conflicts derived from ethnic hatreds and competition for jobs.[15] What is more, environmental changes can lead to humanitarian disasters to which the United States may feel compelled by domestic and world opinion to respond. Such untended environmental threats will leave an unwanted legacy to American children in the form of a polluted, likely hotter, less predictable global environment populated by fewer desirable species. In 1998 Secretary Albright said that environmental threats facing the United States "are not as spectacular as those of a terrorist's bomb or missiles. But we know that the health of our families will be affected by the health of the global environment. The prosperity of our families will be affected by whether other nations develop in sustainable ways. . . .And the security of our nation will be affected by whether we are able to prevent conflicts from arising over scarce resources."[16]

The Clinton administration gave strong *rhetorical* support to a reinterpretation of the national interest that includes environmental change and sustainable development. It showed that it was aware that consumption, not just human numbers, was a major environmental problem. President Clinton, Vice President Gore, and other administration officials acknowledged that Americans had an obligation to reduce their disproportionately high rate of consumption (and, by implication, pollution). Timothy Wirth, then under secretary of state for global affairs, told an audience that the security of the United States hinged "upon whether we can strike a sustainable, equitable balance between human numbers and the planet's capacity to support life." He added that U.S. security "depends on more than military might [because] boundaries are porous; environmental devastation and disease do not stop at national borders." According to Wirth:

The primary threats to human security may not be as easy to recognize as, say, an enemy's nuclear arsenal, but they are no less deadly. These are the threats posed by the abject poverty in which one billion of the world's people live; the hunger that stalks 800 million men, women, and children; the spread of HIV/AIDS. . .; and the combination of violence, poverty, and environmental degradation that have forced 20 million people from their homes. . . . Crises prevention and the challenge of sustainable development are among the great challenges for the remainder of [the twentieth century] and into the next century.[17]

At least by the time of Clinton administration, U.S. officials recognized that successfully combating problems caused by environmental changes required a redefinition and re-prioritization of national interests. To be sure, threats from nuclear proliferation, terrorism, drugs, industrial espionage and the like will and should remain high on the political and national security agenda. But it may be essential for the U.S. government to include environmental changes alongside these more traditional concerns before the processes that are causing the environmental threats become so irreversible that no amount of U.S. effort will be enough to prevent their adverse consequences.

Implementing International Environmental Equity: An Instrument to Defend U.S. National Interests

Under President Clinton, the U.S. government took steps toward a redefinition of which issues constitute core security threats.[18] Policymakers were then faced with finding effective *means* to address environmental threats to U.S. interests. In conjunction with defining adverse environmental changes as first-order threats, the U.S. government had to look for cost-effective and efficacious ways to mitigate or reverse those threats. As suggested previously, this requires the involvement of most of the world's major countries, especially the large developing countries like China, Brazil, and India that are rapidly outpacing the developed countries as the primary contributors to adverse global environmental changes. Indeed, it is precisely these countries that the United States wants to join commitments to one day limit their greenhouse gases in the context of the climate change negotiations.[19]

Developing countries will be more likely to participate in international environmental institutions if they believe that they are being treated with fairness and equity. This means giving due consideration to the historical consequences of industrialization, to the world's skewed consumption patterns, and to developed-country capacities to finance the restructuring that will be necessary to achieve sustainable development and thereby mitigate threatening environmental change. As Philip Shabecoff reminds us, in addition to highlighting the vulnerability of the global environment, international environmental deliberations like those at the Earth Summit "drove home the lesson that unless we care for human needs and wants, we cannot preserve the habitability of the planet. [T]he imminent threats to security were recognized to be the twinned crises of economic inequity and ecological destruction."[20]

Alongside a re-conceptualization of core threats to the national interest, therefore, the requirements of international environmental cooperation to address global environmental changes highlight the need for a greater emphasis in U.S. foreign policy on means beyond traditional military and economic resources for protecting and advancing U.S. interests. Ideas, reputation and the capacity to shape institutions are forms of "power" that can be converted into achievement of U.S. objectives in international affairs. As Joseph Nye points out, if a state "can help support institutions that encourage other states to channel or limit their activities in ways the dominant state prefers, it may not need as many costly exercises of coercive or hard power in bargaining situations."[21]

The United States is likely to retain its superpower status in terms of traditional power resources for decades to come. However, as Nye argues, a country's power is not found in simply material resources but in the extent to which it is able to change the behavior of other countries. "Thus the critical question for the future United States is not whether it will start the [twenty-first] century as a superpower with the largest supply of resources, but to what extent it will be able to control the political environment and to get other nations to do what it wants. The tasks involved in maintaining superpower status will become more complicated in the coming decades, involving a far broader range of issues and a wider variety of players."[22] American "coercive" or "hard" power—a large economy and vast military arsenal—frequently cannot be converted into desired behavior by other countries and indeed by non-state actors. Witness the intransigence of developing countries at the Kyoto climate change conference or difficulties associated with managing the Asian economic crisis of the late 1990s.

To maintain its power, the United States should find methods to persuade countries to behave in ways that promote an orderly international system conducive to its interests, and that simultaneously give other countries reason to believe that their own interests will be promoted.[23] There is, according to Nye, "an indirect way to exercise power."

A country may achieve the outcomes it prefers in world politics because

> other countries want to follow it or have agreed to a system that produces such effects. In this sense, it is just as important to set the agenda and structure the situations in world politics as it is to get others to change in particular situations. This aspect of power—that is, getting others to do what you want—might be called indirect or co-optive power behavior. It is in contrast to the active command power of getting others to do what you want. Co-optive power can rest on the attraction of one's ideas or on the ability to set the political agenda in a way that shapes the preferences that others express.[24]

This "soft" co-optive power complements more traditional forms of command power. Not surprisingly, "If a state can make its power legitimate in the eyes of others, it will encounter less resistance to its wishes."[25]

However, the United States seems to seldom recognize—or at least seldom act on the recognition—that a country can bolster its long-term influence by establishing a reputation as an equitable actor. Much as people assume that other individuals will behave in ways that are similar to their past behavior, rational governments will likely expect other countries to act in ways similar to their past actions. As expressed by Robert Keohane, if "a continuing series of issues is expected to arise in the future, and as long as actors monitor each other's behavior and discount the value of agreements on the basis of past compliance, having a good reputation is valuable even to the egoist whose role in collective activity is so small that she would bear few of the costs of her own malefactions."[26] Put more simply, "A good reputation makes it easier for a government to enter into advantageous international agreements; tarnishing that reputation imposes costs by making agreements more difficult to reach."[27]

Keohane points out that morality can "pay" for governments. By adhering to accepted moral codes, states can identify themselves as "political cooperators" who are among the group of actors "with whom mutually beneficial agreements can be made." According to Keohane,

> publicly accepting a set of principles as morally binding may perform a labeling function. If the code were too passive—turn the other cheek—the moralist could be exploited by the egoist, but if the code prescribes reciprocity in a "tit-for-tat" manner, it may be a valuable label for its adherents. Each egoistic government could privately dismiss moral scruples, but if a moral code based on reciprocity were widely professed it would be advantageous for even those governments to behave as if they believed it. Vice would pay homage to virtue.[28]

Robert McElroy has shown that "equitable trust" forms a basis for the observance of moral norms in the international system.[29] States find that they can benefit by entering into continuing cooperative relationships even when the exact obligations and responsibilities of participants cannot be specified in advance. But in such arrangements states want to ensure, as much as they can, that other participants will behave equitably in those unspecified situations. Consequently, countries with reputations for equitable conduct will substantially enhance their abilities, first, to enter into cooperative relationships, and, subsequently, to profit from those relationships. Importantly, the advantages that accrue for those states that have developed reputations for acting equitably are not restricted to relations between states with similar power capabilities: even "a hegemon will have an interest in developing a reputation for behaving equitably in its cooperative relationships so that it can form such relationships with minimal compliance costs."[30] Thus, most or all states have an incentive to acquire the label "equitably trustworthy." Following moral norms can help states do that. Those not following international moral norms may be labeled as "non-equitable," thereby undermining their ability to join and influence subsequent cooperative arrangements—and also undermining the ability of hegemons to gain cost-effective compliance with those arrangements. McElroy argues that:

> It is not really surprising that moral norms should serve this function. [Citing Ian Macneil,[31]] "In ongoing contractual relations we find such broad norms as distributive justice, human dignity, social equality and inequality, and procedural justice." The fact that a nation follows an international moral norm can signal that the nation respects these values in its international dealings; conversely, the fact that a nation violates an international moral norm can lead other states to label it noncooperative and exploitative.[32]

Such notions can be applied to contemporary security issues. Lawrence Freedman argues that Western democracies have more discretion when engaging contemporary security problems. "Credible rationales" are now critical to building support—both domestically and internationally—for security actions. These new security imperatives have, according to Freedman, "encouraged sensitivity to the normative [and] ethical dimensions of foreign policy."[33]

Many countries now seem to be seriously talking about equity, much more so than during the heyday of the NIEO—at least in the environmental issue area. Among the noteworthy examples of this are the equity provisions of recent international environmental agreements, such as the Montreal Protocol on stratospheric ozone depletion and the climate change and biodiversity conventions signed at the Earth Summit. Those instruments demand that developed countries provide new and additional funds to help developing countries meet the provisions of the agreements. They call for concessional technology transfers from North to South, and they make adjustments to funding arrangements to allow the least affluent parties to take part in decisions regarding the allocation of funding to address adverse global environmental changes.

Once equity becomes a common part of international discourse, it becomes a norm and discussions turn to the subject of how to operate in that norm. Equity takes on a somewhat independent role that is beyond traditional power relationships and calculations. According to Levy, Keohane and Haas, "When international principles and norms have been agreed upon, they may acquire a certain legitimacy and come to be regarded as premises, or as intrinsically valuable, rather than as contestable reflections of interest-based compromises."[34] Working Group III of the Intergovernmental Panel on Climate Change observed that "A broad view of self-interest often points towards explicit consideration of equitable outcomes because of the longer-term risks that grossly inequitable behavior may pose to stability and cooperation in the international system."[35] Increasingly, it is necessary to sell one's objectives to others based on fairness. This suggests that the contexts of international dialogue are changing, as are the frameworks underlying many international organizations and institutions. It also suggests a change in the way countries behave toward one another, albeit only marginally so far.

The U.S. government was slow to recognize the utility of embracing equity norms and objectives. Many participants in the Earth Summit commented that the United States—clearly the most powerful country on earth—failed to grasp the "historic opportunity to build the foundations of a new, cooperative international regime."[36] As Shabecoff pointed out, then Senator Al Gore said at the summit that "Every nation in the world is looking to the United States for leadership."[37] They did not get it. Commenting from the November 1999 Berlin conference of the parties to the FCCC, Philip Clapp, president of the U.S.-based National Environmental Trust, noted that the United States had not changed its ways. It continued to resist calls for it to reduce its greenhouse gas emissions: "If it continues to isolate itself from the rest of the world community by insisting on having its way at every juncture," according to Clapp, "Americans may actually get their wish. One day, they may discover that its posturing has alienated the United States from its allies, angered its trading partners, and jeopardized its leadership role in the world."[38] This came close to happening during the 2000 conference of the FCCC parties in The Hague. The advent of Clinton and Gore to the White House, while leading to a shift in policy toward modest acceptance of equity goals, did not lead to a strong *embrace* of equity as a policy goal in the environmental issue area—let alone in international policy more broadly.

However, the United States may be able to establish (or maintain) itself as a legitimate leader of the world in the twenty-first century by showing that it supports policies intended to bolster international environmental equity, and by showing that it supports equitable arrangements in international affairs more generally. The United States can simultaneously establish a reputation as an equitable partner in international environmental politics and promote its interests in preventing or limiting environmental change. Many of the West European governments recognized the usefulness of being seen as leaders on global environmental issues, as demonstrated by their positions during UNCED and subsequent international environmental meetings. More than the United States, they were and are more sympathetic to the reasonable demands for equity coming from the developing countries, and presumably they understand the utility of such an attitude.

Given a successful outcome on this more limited environmental agenda, the United States might then vigorously promote considerations of international equity in international relations generally, thereby establishing its reputation as not only the most "powerful" country in the world, but also one that all states can expect to deal with them fairly and equitably. While such a policy would be a substantially new one for the United States, and would be costly in terms of the financial and diplomatic resources it would require, at least compared to current meager outlays, in the long run it would likely be less costly and more conducive to the promotion of U.S. global interests than is present policy. For starters, the United States would have to substantially increase its foreign aid spending, and particularly increase its sustainable development aid to the poor countries.[39]

Perhaps what is needed in the medium-term is "enlightened" U.S. leadership—dare one use the word hegemony?—to prod the world in a direction that does less harm to the ecosystem on which all states and their citizens depend for survival.[40] (From time to time, there is much interest within American academic and policy circles in the merits of U.S. global hegemony.[41]) However, such a policy would have to be based in large part on principles of international equity if U.S. leadership were not to be resisted by other countries.

International Environmental Equity and U.S. Foreign Policy: Some Normative Implications

In addition to promoting U.S. global interests, a more robust acceptance by the U.S. government of international equity as an objective of global environmental policy—and indeed foreign policy generally—has potentially beneficial implications for humankind. Implementation of the equity provisions of international environmental arrangements would reduce human suffering by helping to prevent changes to local, regional and global environmental commons that adversely affect people, most notably the many poor people in the developing countries who are least able to cope with environmental changes. Insofar as environmental protection policies focus on sustainable economic development, human suffering may be mitigated as developing countries—especially the least developed countries—are aided in meeting the basic needs of their citizens. Economic disparities within and between countries are growing. At least one-fifth of the world's population already lives in the squalor of absolute poverty.[42] This situation can be expected to worsen in future. If this process can be mitigated or reversed by international policies focusing on environmentally sustainable economic development, human well-being on a global scale will rise.

What is more, international cooperative efforts to protect the environment that are made more likely and more effective by provisions for international equity, if they are successful, will help governments protect their own *and* the global environment. Insofar as the planet is one biosphere—that it is in the case of ozone depletion and climate change seems indisputable—persons in every local and national community are simultaneously members of an interdependent whole. Most activities, especially widespread activities in the United States and the rest of the industrialized world, including the release of ozone-destroying chemicals and greenhouse gases, are likely to adversely affect many or possibly all persons on the planet. Efforts to prevent such harm or make amends for historical harm (i.e., past pollution, which is especially important in these examples because many pollutants continue doing harm for years and often decades) require that most communities work together. Indeed, affluent lifestyles in the United States, Western Europe and other developed areas may harm people in poor areas of the world *more* than those enjoying such lifestyles, because the poor are ill equipped to deal with the consequences.[43] Furthermore, by concerning themselves with the consequences of their actions on the global poor and polluted, Americans and the citizens of other developed countries will be helping their immediate neighbors—and themselves—in the long run. Actualization of international equity in conjunction with sustainable development may help prevent damage to the natural environment worldwide, thereby promoting human prosperity.

The upshot is that the United States has not gone far enough in actively accepting equity as an objective of global environmental policy. It ought to go further in doing so for purely self-interested reasons. However, there are more than self-interested reasons for the United States to move in this direction. It ought to embrace international equity as an objective of its global environmental policy for ethical reasons as well. As argued in chapter 2, we can find substantial ethical justification for the United States, in concert with other developed countries, supporting politically and financially the codification and implementation of international equity considerations in international environmental agreements. Invoking themes found in the corpus of ethical philosophy (but without here assuming the burden of philosophical exegesis!), the United States ought to adopt policies that engender international equity in at least the environmental field (1) to protect the health and well-being of the human species; (2) to universally promote basic human rights; (3) to help the poor be their own moral agents (a Kantian rationale); (4) to help right past wrongs and to take responsibility for past injustices (i.e., past and indeed ongoing U.S. pollution of the global environment); (5) to aid the world's least advantaged people and countries (a Rawlsian-like conception); (6) and because impartiality requires it (among other ethical reasons)—all in addition to the more clearly self-interested justifications that doing so will bolster U.S. credibility and influence in international environmental negotiations and contemporary global politics more generally.

One might argue, therefore, that the United States ought to be aiding the developing countries to achieve sustainable development because to do so may simultaneously reduce human suffering and reduce—and potentially reverse—environmental destruction that could otherwise threaten the healthy survival of the human species. Insofar as human-caused pollution and resource exploitation deny individuals and their communities the capacity to survive in a healthy condition, the United States, which consumes vastly more than necessary, has an obligation to stop that unnecessary consumption. From this basic rights perspective,[44] the U.S. government should also take steps to substantially reduce emissions of pollutants from within the United States that harm people in other countries.[45] The United States ought to refrain from unsustainable use of natural resources or from pollution of environmental commons shared by people living in other countries—or at least make a good effort toward that end—because the people affected by these activities cannot reasonably be expected to support them (we would not be treating them as independent moral agents, to make a Kantian argument[46]). The developed countries, especially the United States, deserve the bulk of the blame for historical pollution contributing to climate change. While the developing countries' net emissions of greenhouse gases are growing fast, their per capita emissions will remain very low relative to the United States, Australia, and other developed countries.[47] Thus the United States ought to make sure, using Henry Shue's words, that "Poor nations ought not be asked to sacrifice in any way the pace or extent of their own economic development in order to help to prevent the climate [or other environmental] changes set in motion by the process of industrialization that has enriched others [i.e., Americans]."[48] The United States also ought to undertake policies that help the least well-off countries and individuals in the world who are suffering from environmental changes (e.g., the people of Bangladesh and the small island states who will suffer—they argue that they are already suffering— the most from sea-level rise caused by climate change), because that is likely what Americans would expect of others were the United States in similar circumstances.[49]

Furthermore, the United States ought to enter international environmental negotiations with the expectation that it should be impartial in its treatment of others, at least insofar as vital American interests are not jeopardized by doing so. American statespersons ought to ask what is reasonable behavior of their government. Is it reasonable to continue to emit pollutants that contribute to climate change and that will likely have especially adverse effects in the poorest countries, and is it reasonable to deny developing countries the help they require to join in efforts to prevent climate change, especially when the United States is disproportionately at fault? The answer seems to be that it is not, leading one to conclude that the United States ought to join international agreements that place requirements on its people to change such unreasonable situations in favor of those adversely affected. In short (repeating a quote from chapter 2), "The United States should aid Bangladesh [and other countries threatened or harmed by climate change] not because it is in the United States' interest to do so but because justice as impartiality suggests that the case for such aid cannot be reasonably denied."[50]

Even Hans Morgenthau, perhaps the most important "realist" theorist of international relations in the twentieth century, thought that morality could and should influence the foreign policies of nation-states, although he saw it as being much less significant than the garnering of traditional power resources to ensure national security. As he observed in *Politics Among Nations,* if we ask ourselves what statesmen and diplomats are capable of doing to further the power objectives of their respective nations and what they actually do, we realize that they do less than they probably could and less than they actually did in other periods of history. They refuse to consider certain ends and to use certain means, either altogether or under certain conditions, not because in the light of expediency they appear impractical or unwise but because certain moral rules interpose an absolute barrier. Moral rules do not permit certain policies to be considered at all from the point of view of expediency. Certain things are not being done on moral grounds, even though it would be expedient to do them. Such ethical inhibitions operate in our time on different levels with different effectiveness. Their restraining function is most obvious and most effective in affirming the sacredness of human life in times of peace.[51]

When a country's vital interests would not be harmed by doing so, it can and should act to promote international moral norms—or at least not act in defiance of them.[52] As McElroy has argued, when the security of a country would be jeopardized by adherence to an international moral norm, the existence of that norm will have little effect on state actions. However, in the day-to-day processes of foreign policy decision making, where the economic and military security of the country are not immediately endangered by compliance with the international moral norm, "the existence of such a norm can prove decisive in determining state behavior."[53] Similarly, as argued by former U.S. national security advisor Zbigniew Brzezinski, "In a world of fanatical certitudes, morality could be seen as redundant; but in a world of contingency, moral imperatives then become the central, and even the only, source of reassurance. Recognition of both the complexity and the contingency of the human condition thus underlines the *political* need for shared moral consensus in the increasingly congested and intimate world of the twenty-first century."[54] In short, when relative security affords statespersons the freedom to choose between a variety of policy choices (which is the normal condition for the United States today), there will be substantial opportunities for morality to shape major foreign policy decisions.[55]

Normatively pregnant ideas can and should influence state behavior. Without the overarching threat of the Soviet Union and the perceived menace of global communism, and without any major country or force to replace them, moral principles that go beyond immediate prudential interests may become more important. In the present post-cold war context, where knee-jerk advocacy of security at the expense of ethics seems especially out of place, morality and ethical imperatives can and probably should influence foreign policy. Especially in international politics surrounding the environmental issues that have become more salient in the wake of the cold war, notably climate change, considerations of international ethics interact with conceptions of national interests. One lesson seems clear: by acting in its self-interest to limit adverse global environmental change, the United States can often promote ethical causes like international equity, potentially helping people suffering from squalor and environmental degradation in the poor countries of the world. By promoting nascent international moral norms of equity, the United States can promote justice in the world and reduce human suffering, while also promoting its national interests with regard to environmental change and international relations more broadly.

During the administration of President Jimmy Carter, when it was U.S. government policy to actively support and promote individual human rights the world over, albeit selectively, many American policymakers were viewed as unrealistic utopians. But with the end of the cold war (not to mention the frequent suggestions that the human rights provisions of the Helsinki Accords helped bring it about) and the advent of the Clinton administration, human rights concerns were again a prominent component of the foreign policy debate and were reflected in U.S. policy. Policymakers that advocated human rights—especially democratic rights— were no longer reflexively labeled "utopians." Human rights became part of the national interest. Similarly, international equity is not a utopian notion restricted to the pages of moral treatises. Already, serious consideration of equity in international politics and U.S. foreign policy in particular have been legitimized, at least in the environmental issue area. Governments and policymakers that declare themselves in support of international equity are not viewed as heretics that fail to understand the dynamics of the anarchic international system. In the real-world context of increasingly global environmental change, promoting international equity can promote national interests.

Conclusion

A decade and a half ago, George Kennan, the former U.S. diplomat who coined the notion of "containment" that was adapted by policymakers to guide U.S. cold war foreign policy—and one who is hardly inclined toward moral sentimentalism—noted that the world is faced with two unprecedented and "supreme" dangers: major war among the great powers and "the devastating effect of modern industrialization and overpopulation on the world's natural environment." According to Kennan:

> The need for giving priority to the averting of these two overriding dangers has a purely rational basis—a basis in national direct interest—quite aside from morality. For short of nuclear war, the worst that our Soviet rivals could do to us, even in our wildest worst-case imagining, would be a far smaller tragedy than that which would assuredly confront us (and if not us, then our children) if we failed to face up to these two apocalyptic dangers in good time. But is there not also a moral component to this necessity? . . . Is there not a moral obligation to recognize in this very uniqueness of the habitat and nature of man the greatest of our moral responsibilities, of ourselves, in our national personification, its guardians and protectors rather than its destroyer?[56]

The United States arguably has an obligation as the sole superpower, as the wealthiest country in the world, and as the greatest polluter on the planet (among other reasons) to become a leader in efforts to promote international environmental equity. A fairer and more just sharing among the world's countries of the benefits and burdens of international environmental relations like the climate change agreements will contribute to meeting many of the reasonable demands for equity from the world's poor and weak countries and people. Managed in an efficacious manner, such a redistribution of benefits and burdens could simultaneously fulfil ethical objectives, reduce human suffering (and the suffering of other species), and protect the local, regional and global natural environments.

These measures would also promote American national interests by reducing the potential adverse economic and security consequences for the United States of environmental change and poverty in the developing world. This type of foreign policy would bolster the reputation and credibility of the United States, meaning that other countries would be more likely to turn to it for leadership and to trust it in international deliberations in other issue areas. Using the words of former President Jimmy Carter, "we should take on this global leadership role, both because it is right and because it is in our self-interest to do it."[57] In short, by doing what is right—by doing what is equitable—in its relations with the rest of the world, the United States can best promote its national interests and ensure its future status as a great power.

Notes

1   The absurdity of this example shows the limited utility of weapons of mass destruction—weapons that are, strictly speaking, the most "powerful"—in almost all day-to-day issue areas.

2   I also made this argument in Paul G. Harris, "Environment, Security and Human Suffering: What Should the United States Do?," in Paul G. Harris, ed., *The Environment, International Relations, and U.S. Foreign Policy* (Washington: Georgetown University Press, 2001), pp. 241-266.

3   This was one of the messages of the famous Brundtland Report and General Assembly resolutions underlying the 1992 United Conference on Environment and Development (UNCED). See World Commission on Environment and Development (Brundtland Commission), *Our Common Future* (Oxford: Oxford University Press, 1987); United Nations General Assembly Resolution 44/228, "United Nations Conference on Environment and Development," A/44/228, 22 December 1989.

4   In addition to chapters 3 and 4, also James Cameron and Ruth MacKenzie, "Access to Environmental Justice and Procedural Rights in International Institutions," in Alan Boyle and Michael Anderson, *Human Rights Approaches to Environmental Protection* (Oxford: Clarendon Press, 1998), pp. 129-52.

5   In 1997, U.S. non-military foreign aid was only $7 billion, less than one tenth of one percent of GNP, and half what it was ten years earlier. Karen DeYoung, "U.S. Grows Stingier on Foreign Aid," *International Herald Tribune*, 26 November 1999, p. 1.

6   See Paul G. Harris, "Considerations of Equity and International Environmental Institutions," *Environmental Politics* 5, 2 (Summer 1996), pp. 274-301.

7   Many developed countries, the United States included, have been slow to recognize the links between well-being, poverty, crime and environmental degradation in their own backyards. However, the Environmental Protection Agency has taken notice, and it now has a program devoted to environmental justice issues. See Robert D. Buller, ed., *Unequal Protection: Environmental Justice and Communities of Color* (San Francisco: Sierra Club Books, 1994); Richard Hofrichter, *Toxic Struggles: The Theory and Practice of Environmental Justice* (Philadelphia: New Society Publishers, 1993); and Evan J. Ringquist, "Environmental Justice: Normative Concerns and Empirical Evidence," in Vig and Kraft, pp. 232-256 (and works cited therein).

8   See Montreal Protocol on Substances that Deplete the Ozone Layer, 16 September 1997, 26 *International Legal Materials* 1550 (1987); United Nations Framework Convention on Climate Change (UN FCCC), 9 May 1992. See also chapter 4 and Jay Shulkin and Paul Kleindorfer, "Equity Decisions: Economic Development and Environmental Prudence," *Human Rights Quarterly* 17 (1995), pp. 382-97.

9   The connections between environmental degradation and poverty are described in a vast literature. See particularly the Brundtland Report.

10   For a discussion, see Judy Mann, "Reviewing History's Lesson on Foreign Aid," *Washington Post* (1 July 1998).

11   Thomas Homer-Dixon, "On the Threshold: Environmental Changes as Causes of Acute Conflict," *International Security* 16, 2 (1991), pp. 76-116; Norman Myers, *Ultimate Security: The Environmental Basis of Political Stability* (New York: W.W. Norton, 1993).

12   Warren Christopher, "Foreign Affairs Budget that Promotes U.S. Interests," statement before the Subcommittee on Foreign Operations of the Senate Appropriations Committee, Washington, DC, 2 March 1994, *U.S. Department of State Dispatch* 5, 11 (14 March 1994).

13 Ibid.

14 Department of State, *Environmental Diplomacy: The Environment and U.S. Foreign Policy* (Washington: Department of State, 1996), <http://www.state.gov/www/global/oes/earth.html>.

15 Alan Dowty and Gil Loescher, "Refugee Flows as Grounds for International Action," *International Security* 21, 1 (1996), pp. 43-71; Evan Vlachos, "Environmental Threats and Mass Migrations: The Growing Challenge of Environmental Refugees," paper presented at the NATO Advanced Research Workshop, Bolkesjo, Norway, 12-16 June 1996; Myron Weiner, "Bad Neighbors: An Inquiry into the Causes of Refugee Flows," *International Security*, 21, 1 (1996), pp. 5-42.

16 Madeleine Albright, "Earth Day 1998: Global Problems and Global Solutions," Press Release, Department of State, 21 April 1998, <http://secretary.state.gov/www/statements/1998/980421.html>.

17 Timothy E. Wirth, "Sustainable Development and National Security," address before the National Press Club, Washington, DC, 12 July 1994, *U.S. Department of State Dispatch* 5, 30 (25 July 1994).

18 National Security Council, *A National Security Strategy for a New Century.* (Washington: National Security Council, May 1997).

19 Paul G. Harris, "Les Etats-Unis: Un Joueur Cle," *Liaison Energy Francophonie* 39 (1998), pp. 11-15.

20 Philip Shabecoff, *A New Name for Peace: International Environmentalism, Sustainable Development, and Democracy* (Hanover, NH: University Press of New England, 1996), pp. 176-77.

21 Joseph S. Nye, Jr., *Bound To Lead: The Changing Nature of American Power* (New York: Basic Books, 1990), p. 33. Nye subsequently went to work for the Pentagon during the Clinton administration.

22 Ibid., p. 175.

23 Cf. Robert W. Cox, *Production, Power, and World Order* (New York: Columbia University Press, 1987).

24 Nye, p. 31.

25 Ibid., p. 32.

26 Robert Keohane, *After Hegemony* (Princeton: Princeton University Press, 1984), pp. 105-106.

27 Ibid, p. 106.

28 Ibid., p. 127.

29 Robert W. McElroy, *Morality and American Foreign Policy* (Princeton: Princeton University Press, 1992), p. 48.

30 Ibid., p. 49.

31 Ian Macneil, "Contracts: Adjustments of Long-Term Economic Relationships under Classical, Neoclassical, and Relational Contract Law," *Northwestern University Law Review* 72 (1977-78), p. 898.

32 McElroy, p. 52.

33 Lawrence Freedman, "International Security: Changing Targets," *Foreign Policy* 110 (Spring 1998), p. 58.

34 Marc A. Levy, Robert O. Keohane, and Peter M. Haas, "Improving the Effectiveness of International Environmental Institutions," in Peter M. Haas, Robert O. Keohane and Marc A. Levy, *Institutions for the Earth* (Cambridge, MA: MIT Press, 1993), p. 400.

35 Tariq Banuri et al., *Intergovernmental Panel on Climate Change, Working Group III: Equity and Social Considerations*, draft 8.5, mimeo, 1995, pp. 4-5.
36 Shabecoff, p. 166 (based on Shabecoff's first-hand observation of the Earth Summit).
37 Ibid., p. 167 (based on Shabecoff's interview with Gore).
38 Philip E. Clapp, "America Must Act on Global Warming," *International Herald Tribune*, 4 November 1999, p. 9.
39 Carol Lancaster, "Redesigning Foreign Aid," *Foreign Affairs* 79, 5 (September/October 2000), pp. 74-88.
40 See John Barkdull and Paul G. Harris, "The Land Ethic: A New Philosophy for International Relations," *Ethics and International Affairs* 12 (1998), pp. 159-77.
41 See, for example, William Pfaff, "America in History: Realists Don't Buy the Wilson Line," *International Herald Tribune* (8 July 1998).
42 See recent issues of the World Bank's *World Development Report* (Washington: World Bank, annual).
43 See, for example, Robert T. Watson, Marufu C. Zinyowera, Richard H. Moss, eds., *The Regional Impacts of Climate Change: An Assessment of Vulnerability* (Geneva: Intergovernmental Panel on Climate Change, November 1997).
44 Cf. Henry Shue, *Basic Rights*, 2nd ed. (Princeton: Princeton University Press, 1996). See generally Boyle and Anderson.
45 I am not suggesting that Americans ought to reduce their standard of living. Rather, they should live the good life in a more sustainable way, something modern technologies make possible. On the potential of such technologies, see for example Amory Lovins and Hunter Lovins, "Climate Change: Making Sense and Making Money," White Paper of the Rocky Mountain Institute (1997).
46 Cf. Onora O'Neill, *Faces of Hunger: An Essay of Poverty, Development and Justice* (London: Allen and Unwin, 1986). See Onora O'Neill, "Kantian Ethics," in Peter Singer, ed., *A Companion to Ethics* (Oxford: Blackwell, 1993), pp. 175-85.
47 An anonymous reader points out that this important issue is illustrated by the emissions of India: In a year, all of India uses about the same amount of energy as greater Los Angeles (!).
48 Henry Shue, "The Unavoidability of Justice," in Andrew Hurrell and Benedict Kingsbury, eds., *The International Politics of the Environment* (New York: Oxford University Press, 1992), p. 395.
49 This is a Rawlsian argument. Cf. John Rawls, *A Theory of Justice* (Cambridge, MA: Harvard University Press, 1971).
50 Borrowing the words of Chris Brown used in general reference to impartiality regarding international distributive justice. Brown, *International Relations Theory*, pp. 180-81. This is an argument based on impartiality. Cf. Brian Barry, *Theories of Justice* (Berkeley: University of California Press, 1989) and Barry, *Justice as Impartiality* (Oxford: Clarendon Press, 1995).
51 Hans Morganthau (6th ed. revised by Kenneth W. Thompson), *Politics Among Nations* (New York: McGraw-Hill, 1985), p. 249.
52 Cf. Arnold Wolfers, *Discord and Collaboration* (Baltimore: Johns Hopkins University Press, 1962) in which Wolfers describes the spectrum of pressures influencing policymakers. On one hand is the "pole of necessity" where the state's national security will be promoted by one choice and harmed by all others. On the other hand is the "pole of choice" where the state's security is not threatened or where several possible policy choices may promote the state's security.
53 McElroy, p. 183.

54 Zbigniew Brzezinski, *Out of Control: Global Turmoil on the Eve of the Twenty-first Century* (New York: Charles Scribner's Sons, 1993), p. 231.
55 McElroy, p. 184.
56 George Kennan, "Morality and Foreign Policy," *Foreign Affairs* 64, 2 (Winter 1985/86), p. 217.
57 U.S., Congress, House of Representatives, Committee on Foreign Affairs, Subcommittee on Western Hemisphere Affairs, statement of President Jimmy Carter, *The United Nations Conference on Environment and Development*, 102nd Cong. (4 February 1992), 1993, p. 7.

10 Conclusion: International Environmental Equity in U.S. Foreign Policy

The previous chapters have attempted to answer the following primary questions:

- To what extent has the U.S. government—and indeed other developed countries—accepted international equity as an objective of global environmental policy?
- What explains the U.S. government's limited acceptance of international equity as an objective of its policy in the global environmental field?
- Why did the U.S. government under President Bush not go further in accepting international equity as an objective of U.S. global environmental policy? Why did the Clinton administration go beyond the Bush administration in this regard? Why did it not do more to implement this policy?
- Should the United States (and other developed countries) go further to *embrace* international environmental equity as an objective of global environmental policy?

The Emergence and Acceptance of International Equity in Global Environmental Politics

There is ample historical evidence to show that international equity has started to permeate global environmental politics. Since the Stockholm Conference thirty years ago, the developed countries have come to see international equity as an unavoidable and even desirable component of efforts to protect the world's natural environment. Equity was reflected in international environmental instruments agreed during the 1980s, such as the Law of the Sea and the Montreal Protocol on Substances that Deplete the Ozone Layer. The most extensive manifestations of the emergence of international environmental equity were the process of the United Nations

244

Conference on Environment and Development (UNCED), the agreements and conventions signed at the 1992 Rio Earth Summit, and the follow-on deliberations regarding those agreements and treaties. Provisions for international environmental equity in the UNCED agreements included calls for new and additional funds and concessional technology transfers to help developing countries and particular populations in those countries develop in a sustainable fashion and meet the provisions of UNCED agreements and conventions. Examples in the follow-on agreements included changes to international environmental funding institutions giving developing countries greater authority in decisions regarding the allocation of financial assistance, and the Kyoto Protocol to the Climate Change Convention, which required developed countries—and specifically *not* developing ones—to reduce their emissions of pollutants that contribute to global climate change.

Prominent emerging sub-themes of international environmental equity include: recognition by most countries that the North is inordinately the cause of historic global environmental pollution and its contemporaneous global effects, and that the North is obligated to more aggressively address that situation due to its responsibility for the pollution and its greater ability to pay for action; the priority of economic development in the South and the North's recognition and acceptance that the developing countries ought not be diverted from economic development by environmental protection measures; recognition that new and additional funds (over and above prevailing levels of official development assistance) will be required for sustainable development programs in the poor countries; agreement that technology transfer on preferential terms will be required if developing countries are to protect their own and the global environment; developing countries rightly own and control resources within their national jurisdictions and should benefit from products developed from those resources; developing countries who are recipients of aid for sustainable development should participate in the decision making of international environmental funding instruments; sustainable development—the integration of environmental protection, economic development and poverty eradication—should be the guiding theme of international and especially North-South environmental policies.

The United States slowly joined the emerging consensus, although it did so to a degree that leaves very much to be desired from the perspective of those who support robust efforts to promote international equity and global environmental protection. In the period between the start of the Bush administration and the end of the Clinton administration, the U.S. government's policy on international environmental equity shifted from outright opposition to partial embrace. During the first three years of the Bush administration, U.S. policy toward international environmental equity was one of disinterest followed by adamant opposition. When policymakers took an interest in this issue they used U.S. diplomatic influence to oppose measures that would set precedents for equity in international environmental agreements. In the last year of the Bush administration, however, there was a small but noticeable shift in policy toward marginal acceptance or at least lack of opposition to the idea of international environmental equity. The U.S. government was either unable or unwilling to oppose many efforts by American and foreign interests, including other governments, to incorporate considerations of equity into ongoing UNCED negotiations and the agreements and treaties they produced. Thus, international equity provisions permeated the international instruments agreed during UNCED and signed at the Earth Summit. After the Earth Summit the Bush administration was less obstructionist in international environmental deliberations than it was before 1992, and it supported, at least rhetorically, most of the UNCED agreements. To be sure, those agreements contained watered-down equity provisions, but those provisions set precedents for the future that were constantly referred to by actors—politicians, nongovernmental organizations, other governments, and other actors—in their dealings with the Bush administration during the last three-quarters of 1992.

Acceptance by the U.S. government of international equity as an objective of global environmental policy was much more evident during the Clinton administration. During the 1992 presidential campaign, Clinton and his vice-presidential running mate Al Gore professed support for the international environmental agreements and conventions that emanated from the UNCED process. For example, Gore acknowledged that Americans and citizens of other developed countries would have to reduce their high levels of consumption (which cause environmental degradation well beyond U.S. borders) before asking the developing countries to undertake expensive sustainable development measures, and the Clinton-Gore campaign explicitly stated support for the agreements signed at the Earth Summit. After taking office the Clinton administration began a transition to policies that were more accepting of international environmental equity, with these efforts being reflected in forthright policy pronouncements by 1996. Sustainable development became a central guiding theme of the Clinton foreign policy, consciously and publicly integrated throughout the entire foreign policy establishment. Perhaps most notable was the administration's July 1996 declaration of support for a binding international agreement to reduce emissions of greenhouse gases.

The Clinton administration continued to make international environmental equity part of U.S. foreign policy, despite the opposition of a Republican-controlled Congress generally hostile to considerations of international equity.

Explaining U.S. Policy (I)

The U.S. government's partial acceptance of international equity as an objective of global environmental policy is explained in part by the recognition of U.S. policymakers that threats to important U.S. interests could not be countered without securing the cooperation of developing countries who have been demanding greater international equity.

A rather solid consensus emerged in the U.S. foreign policy community that environmental changes threaten U.S. environmental, economic, and potentially military interests, as well as the interests of U.S. allies. Moreover, adverse environmental changes in the developing countries could contribute to economic stagnation and prevent economic development there, preventing those countries from purchasing American exports and potentially contributing to Northward migrations of economic refugees. It was understood that environmental changes in the developing countries could also contribute to domestic unrest and potentially violent internal and international conflict, with attendant consequences for U.S. interests. There was general recognition that, most important for U.S. interests in the long term, environmentally unsustainable practices in the developing world will contribute to adverse global environmental changes, such as climate change, loss of species, and stratospheric ozone depletion, and that these types of environmental change could directly affect the U.S. through weather, climate, sea level and other changes within U.S. territory, in addition to exacerbating environmental degradation problems overseas that could affect U.S. interests.

Prevention of global pollution—damage to the global commons—almost by definition requires international cooperation. In the case of climate change, effective measures to reduce emissions of pollutants that cause it require nearly comprehensive international cooperation, and notably require the willing participation of the large developing countries. Neither the United States nor any other country by itself can effectively address these problems. In return for their participation in international cooperative efforts to prevent continued destruction of the natural environment, developing countries have demanded and will continue to demand concessions from the North. Provisions for international equity came to be viewed by American policymakers as a necessary means by which to garner developing country support for international environmental arrangements that were deemed necessary to protect U.S. national interests. Thus U.S. acceptance of international equity as an objective of global environmental policy was in part a result of the major environmental issues themselves. Many of these issues, unlike most other issues in international relations, are ones in which U.S. national interests are directly connected to the actions of the developing countries, making their demands for international equity more salient than ever before.

Explaining U.S. Policy (II)

The U.S. government's partial acceptance of international equity as an objective of global environmental policy was not only a direct response to requirements of the new environmental diplomacy, but was also shaped in large part by the influence of various actors in American politics and policy-making processes.

Both the Bush and Clinton administrations were influenced by the same range of actors. These included business and industry interests, nongovernmental organizations, scientists, members of Congress, the American public, other governments, and individuals, especially individuals inside the government who were in positions that afforded them access and influence in the policy process.

During the Bush administration, commercial interests, such as the fossil fuel, automobile, and pharmaceutical industries, tended to support the administration's posture toward the UNCED agreements. In contrast, during the Clinton administration businesses that were opposed to UNCED follow-on arrangements were frequently at odds with the executive branch of government, although they were countered by a growing group of industries (e.g., environmental technology and insurance industries) supportive of UNCED agreements and related follow-on international deliberations.

Environmental and development nongovernmental organizations were supportive of efforts to incorporate international equity into the UNCED process. During the Bush administration, their views were heard by U.S. diplomats, but they had limited effect on leading policymakers. Alternatively, during the Clinton administration, environment and development NGOs received a more sympathetic hearing; indeed, Clinton appointees to several foreign policy posts were drawn from those organizations.

The influence of scientists increased over time. As scientists' ability to more clearly explain and predict the causes and consequences of the greatest environmental threats (especially climate change) increased, they contributed to U.S. partial acceptance of international environmental equity. Policymakers and those outside government interested in having the United States take on many of the burdens of global environmental protection (e.g., reduce U.S. emissions of carbon dioxide and other greenhouse gases) were able to draw greater support from the views of scientists. Thus the second assessment report of the Intergovernmental Panel on Climate Change, which established that climate change and its adverse consequences were in part the consequence of human activities, was eventually endorsed by the Clinton administration and contributed to the administration's agreement to accept a binding international treaty requiring it and other developed countries to reduce their greenhouse gas emissions (i.e., the Kyoto Protocol).

Throughout the UNCED process and follow-on deliberations, coalitions of these and other forces became important. For example, in the case of climate change, during the Bush administration fossil fuel and automobile interests joined forces with a few influential scientific skeptics and the oil producing countries to press the Bush administration to block inclusion of binding greenhouse gas emissions targets and timetables in the climate change treaty. Also during the Bush administration, free-market ideologues within government were able to draw support from the biotechnology industry to prevent U.S. participation in the biodiversity treaty. By contrast, during the Clinton administration, environmental technology and insurance interests cooperating with the bulk of atmospheric scientists and environmental groups pressed for stronger U.S. support of the climate change treaty. Similarly, the Clinton administration signed the biodiversity treaty largely because the biotechnology, pharmaceutical and agricultural industries came out in support of it. In each respective administration, coalitions of actors were generally successful in influencing U.S. policy, gradually nudging the U.S. government toward greater acceptance of international environmental equity.

A crucial part of the explanation for the U.S. government's acceptance of international environmental equity was the influence of individuals in the government. Bush White House chief of staff John Sununu, Vice President Dan Quayle and members of his Council on Competitiveness, Office of Management and Budget director Richard Darman, and Bush political appointees in the State Department (e.g., Under Secretary of State for Policy Affairs Robert Zoellick) were able to block major efforts by U.S. diplomats to incorporate equity into UNCED agreements and especially the conventions. Even so, by the time these anti-environmentalists began to focus on UNCED, the underlying themes of the conference had already been established in international negotiations conducted by lower-level U.S. diplomats and policymakers more sympathetic to international equity objectives. Later in the process, Sununu and others still labored assiduously to limit the international equity provisions of many key agreements. However, they were not able to eliminate those provisions. In contrast, in the Clinton administration, influential individuals were able to move U.S. policy toward a greater acceptance of international environmental equity. Most important was Vice President Al Gore, who came to the administration intent on making global environmental protection a central theme of American foreign policy. His advocacy and the influence of like-minded individuals in the government—whom Gore in effect appointed in some cases—were central to the administration's gradual acceptance of international equity as an objective of global environmental policy.

These actors operated in contexts that affected how they could influence policy. The pluralist nature of American politics and the requirement that the Bush and Clinton administrations go to the polls after four years meant that actors had political access and the ears of both administrations. Bush and Clinton were pulled in directions both critical and supportive of international environmental equity as their re-election campaigns neared; but given the differing makeup of their prime constituencies, the impact on each administration was different. Bush leaned toward those who were opposed—although he could have leaned further. Clinton leaned toward those who were supportive—although his administration was generally long on rhetoric and short on action.

The various actors also operated in the larger context of the emerging global consensus on international environmental equity. Domestic actors and domestic politics were most important, but also important were international institutions in which equity was viewed as a desirable objective of policy and a prerequisite for effective international environmental cooperation. The United Nations and related organizations that shepherded the UNCED agreements and conventions operated with these assumptions, as did many diplomats from other governments and nongovernmental organizations (which were for the first time allowed to participate intimately in international environmental policy making). In short, international environmental equity was one of the guidelines by which the whole UNCED process operated. Although international institutions and the international norms they engendered had an indirect influence on U.S. policy, that influence was important for moving the

United States toward partial acceptance of international equity as an objective of global environmental policy.

Explaining U.S. Policy (III)

The U.S. government's partial acceptance of international equity as an objective of global environmental policy is explained in part by the appeal of international equity as a value in itself.

The idea of international equity appealed to some individuals who were influential in the U.S. foreign policy process. In the case of the Bush administration, some mid-level bureaucrats and diplomats felt that it was their personal obligation and the obligation of the United States to promote the well-being of poor people living in the developing countries. Because they did not see a discrepancy between these sympathies and U.S. national interests—indeed, they tended to view them as entirely complimentary—they were receptive to many of the demands made by the developing countries for international equity considerations in the UNCED process. Especially in the early phases of the UNCED negotiations, these altruistic Americans were able to influence the U.S. posture and hence the entire UNCED process. As UNCED and the associated biodiversity and climate deliberations became more salient to high-level officials of the Bush administration, however, individuals interested in promoting international environmental equity had less influence. Yet, what they had done early on had a lingering effect, and they were again able to have a marginal impact on Bush administration policy in the last months of its tenure.

In the case of the Clinton administration, this process continued to operate, but with greater impact because such sympathies prevailed at the top of the environmental foreign policy process. Vice President Gore viewed protecting the global environment as a moral imperative; he saw equitable relations with the developing countries, sustainable economic development, and poverty eradication as values that could not be separated from this imperative. Furthermore, other people throughout the foreign policy process during the Clinton administration were similarly motivated, many of them having received positions in the government with the help of Gore. Former Senator Timothy Wirth was appointed to a newly created branch of the State Department devoted to sustainable development, human rights and other global issues. Former outspoken environmentalist Rafe Pomerance, who continued to call for concrete U.S. action to curb its greenhouse gas emissions and to help developing countries do likewise, became the Clinton State Department's first point-man on climate change.

There were individuals with altruistic motivations in the government, and they ushered international equity, at least in the context of global environmental policy, through the policy making and implementation process.

Those individuals who viewed international equity as a value worth promoting in its own right operated in—and were no doubt affected by—the larger international milieu of the emerging consensus on international equity and the post-war foreign aid regime. Republican administrations tend to be less willing to provide foreign aid for the purposes of promoting international equity and related objectives; Democratic administrations tend to be relatively more generous in this regard. Altruistic individuals were in hostile circumstances in the Bush administration; like-minded individuals in the Clinton administration were in more sympathetic and receptive circumstances. The foreign aid policies and practices of developed country governments tend to mirror their domestic welfare policies. Governments that provide generous welfare at home tend to be more generous abroad, donating more money per capita, giving more to multilateral institutions, and directing more aid to the poorest countries and people in the world. Governments that are less generous at home tend to give less in foreign development assistance per capita, tend to "tie" most of that aid, and are less likely to give aid to multilateral institutions or to programs designed to meet the basic needs of the world's poor. The U.S. posture with regard to international environmental equity becomes clearer in these historical contexts. The United States was less generous than most other developed countries with regard to international equity in the context of UNCED, but it was also less generous at home. The Clinton administration was more forthcoming than was the Bush administration, much as we would expect of a Democratic administration. Even the fall-off in foreign aid levels after the Earth Summit is made understandable when viewed in light of analogous reductions in domestic welfare spending. But despite such cutbacks, the U.S. government during the Clinton administration endeavored to direct more of its limited resources toward sustainable development in the poorest countries, and at least moved the United States in the direction of taking on a greater share of international environmental burdens.

Conclusion: What Should Be Done?

A more robust acceptance on the part of the U.S. government of international equity as an important goal of global environmental policy faces many obstacles, most notably the pluralistic nature of American domestic politics and foreign policy that permits many disparate and sometimes powerful actors, both domestic and foreign, to influence the policy process. Domestically, business and industry groups, citizens' organizations, scientists and academics, bureaucratic agencies, the media and public opinion, members of Congress and political parties, as well as influential individuals, can work to oppose or support U.S. foreign policies accepting and aiding international environmental equity. Internationally, the same actors in other national jurisdictions, multinational corporations, and international governmental and non-governmental organizations can use the American political process, the international media, and diplomatic channels to influence U.S. policy. As adverse global environmental change becomes more pronounced, rather than largely a *future* threat as it is still often perceived, one would expect many of these forces to give more support—or offer less resistance—to U.S. promotion of international equity as it relates to environmental issues. Already the economically and politically powerful American insurance industry has joined with environmental groups to press the U.S. government to take climate change seriously and to reduce U.S. emissions of carbon dioxide and other greenhouse gases.

But many actors will feel threatened by such policy changes. The American people may remain opposed to more foreign aid because they are very poorly informed about how small it actually is compared to what they expect, and self-interested members of Congress may continue to exploit this ignorance. Many powerful industries will remain opposed to efforts that make the U.S. economy more environmentally benign. Thus, a more robust acceptance by the U.S. government of international equity and burden sharing as an important goal of global environmental policy will require a coalition between sympathetic agencies, diplomats, citizens groups, businesses, and the like. Most important, perhaps, will be strong leadership from the president and executive agencies, in concert with these sympathetic actors, to make clear to the American people and their congressional representatives the connections between U.S. interests and international equity, especially in the environmental field. Successful leadership efforts may also have to draw on the American people's historic willingness to "do good" in the world, especially when doing so can simultaneously promote U.S. interests. Whether the *new* Bush administration will provide this leadership is unclear at the time of this writing, although pessimism seems more than justified.[1]

A full understanding of the U.S. government's partial and belated acceptance of international equity as an objective of global environmental policy, and of its failure to implement these policies more actively, requires considering the role of many state and non-state actors, institutions and organizations, bargaining environments, and the powerful ideas that are salient in contemporary international deliberations on environmental issues—among many other factors.[2] However, while such a complex analysis would help us to better understand what has happened and is happening in international environmental politics, it may not be *sufficient* for nudging practitioners toward formulating and supporting more robust international environmental institutions that aggressively tackle the global environmental problems the world will face in the twenty-first century. What may be needed is a prescription for action that encompasses international equity as an *a priori* principle. To use Shue's formulation, what this means is that "ethical considerations like international justice and human rights must come in at the ground level when notions of legitimate national interests are initially shaped, not as superficial constraints that are essentially after-thoughts."[3] By acting in a narrowly self-interested manner, the United States and other wealthy countries may achieve their short-term aims and, at the same time, do some good. It is also possible, or even likely given the past and future changes to the global environment, that by consciously—and conscientiously—trying to do what is best for the world, these countries will even more effectively and efficiently promote their own interests *and* those of all humankind.

Notes

1   As this book goes to press, President George W. Bush reversed his campaign pledge to place new controls on U.S. carbon dioxide emissions. The White House subsequently announced that the United States would not implement the Kyoto Protocol, and there were indications that the new Bush administration wanted to withdraw the U.S. signature from it. By all indications, the same forces shaping the first Bush administration's policies on climate change were at work on the second one, in precisely the ways described in this book. See Douglas Jehl and Andrew C. Revkin, "President Cancels Clean-Air Vow: No New Carbon Dioxide Limits," *International Herald Tribune*, 15 March 2001; Amy Goldstein and Eric Pianin, "Congressional Pressure Played Role in Bush's Reversal on Emissions," *International Herald Tribune*, 16 March 2001; Eric Pianin and William Drozdiak, "A Setback is Feared on Global Warming: Environmental Groups See Bush's Turnaround as a Bar to World Pact," *International Herald Tribune*, 17-18 March 2001; Associated Press, "U.S. Won't Implement Climate Treaty," *New York Times*, 28 March 2001; Doug Struck, "U.S. Upsets Japan on Ecology Pact: Bid to Ignore Kyoto Accord Opens Rift with Washington," *International Herald Tribune*, 29 March 2001; and Douglas Jehl, "U.S. on Defensive at Ecology Talks: Bush Warns of Harm to Economy," *International Herald Tribune*, 30 March 2001.

2   For detailed analyses of U.S. environmental foreign policies, see Paul G. Harris, ed., *The Environment, International Relations, and U.S. Foreign Policy* (Washington: Georgetown University Press, 2001) and Paul G. Harris, ed., *Climate Change and American Foreign Policy* (New York: St. Martin's Press, 2000).
3   Henry Shue, "Ethics, the Environment, and the Changing International Order," *International Affairs* 71, 3 (1995), p. 461.

Index

About the Author

Paul G. Harris is senior lecturer in international relations at London Guildhall University, where he directs the Project on Environmental Change and Foreign Policy. He is an associate fellow at the Oxford Centre for the Environment, Ethics and Society (OCEES) at Mansfield College, Oxford University, and an honorary research fellow at the Centre of Urban Planning and Environmental Management at the University of Hong Kong. He is also an associate professor at Lingnan University, Hong Kong, where he is a research fellow at the Centre for Asian Pacific Studies and the Centre for Public Policy Studies. Dr. Harris is the editor of, among other books, *Climate Change and American Foreign Policy* (St. Martin's Press/Palgrave) and *The Environment, International Relations, and U.S. Foreign Policy* (Georgetown University Press). He is author of the OCEES report, *Understanding America's Climate Change Policy*, and numerous journal articles and papers on global environmental politics, American foreign policy, and international ethics. He received his Ph.D. from Brandeis University.

Printed and bound by CPI Group (UK) Ltd, Croydon, CR0 4YY

22/10/2024

01777605-0012